The Windsor Locks Canal
Yankee Enterprise and Irish Muscle

THE WINDSOR LOCKS CANAL

The Windsor Locks Canal

Yankee Enterprise and Irish Muscle

by

J. Christopher Kervick

The Connecticut Press
Madison, CT
2025

Copyright © 2025 by J. Christopher Kervick. All rights reserved.

This book may not be reproduced, in whole or in part, in any form *(beyond the copying permitted by Sections 107 and 108 of the U.S. Copyright Law and except by reviewers for the press)*, without written permission from the author.

First printing, 2025

Library of Congress Cataloging in Publication Data
Kervick, J. Christopher, author
The Windsor Locks Canal: Yankee Enterprise and Irish Muscle
Includes illustrations, maps, annotations, bibliography, and index,
198 pps.
ISBN: 978-0-9977907-9-5
Library of Congress Control Number: 2025900966

1. U.S. History - Transportation - Conneticut - Canals
2. Industrial Revolution | Windsor Locks, Connecticut |
3. Social History | Irish Immigrantion - Connecticut Social, Ethnic, and Religious and Demographics

Designed & Printed in

Front Cover*: Colorized version of the 1857 steel engraving of the Persse & Brooks Paper Company of Windsor Locks, CT.*

DEDICATION

To the people of Windsor Locks and Suffield, Connecticicut

CONTENTS

Illustrations		*ix*
Preface		*xi*
Chapter 1	*The Push for Internal Improvements*	*17*
Chapter 2	*Muscle From Abroad*	*27*
Chapter 3	*The Dawn of the Canal Era*	*35*
Chapter 4	*Competition from New Haven*	*41*
Chapter 5	*The River Improvement Scheme*	*61*
Chapter 6	*Locking the Falls*	*81*
Chapter 7	*Building the Canal*	*93*
Chapter 8	*The Irish Labor Force*	*119*
Chapter 9	*Completing the Canal*	*145*
Chapter 10	*Beyond Completion*	*153*
Chapter 11	*Industries Along the Canal*	*169*
Appendix	*Listing of Irish Surnames Associated with the Canal*	*185*
Index		*187*

ILLUSTRATIONS

Head Gate and Guard Lock of Canal	17
The Old Ferry Between Warehouse Point and Windsor Locks, CT	21
Rendition of Irish Family During the Penal Law Era	31
Typical Flatboat Used on River Above Hartford	41
Second Enfield Bridge	43
Depiction of Farmington Canal Boat	55
Portrait of Alfred Smith	67
Steamboat "Barnet"	73
Banknote from The Connecticut River Banking Company	83
Portrait of Chief Engineer Canvass White	93
Section of Canvass White's Canal Plan	95
Father John Power	125
Thomas Blanchard	148
Montgomery Company, Windsor Locks	178
Medlicott Company, Windsor Locks	181
Irish Workers' Memorial, Windsor Locks	186

PREFACE

I grew up in Windsor Locks, Connecticut, three hundred yards from the Windsor Locks Canal. My house was on Elm Street, which is the modern name for the Saw Mill Path used to deliver logs to Dexter's mill. I dug clay from Adds Brook, crossed Kettle Brook on fallen trees, and netted frogs in the Mill Pond fed by both of those streams. I fished in the canal's upper basin, catching stocked trout during the annual Boy Scout fishing derby, and pumpkinseed and bluegill all other times.

My mother worked for the Dexter paper mill for thirty years and I spent five summers working there so I could buy books and beer in college. As a sixteen-year-old, a pretty girl and I took an unauthorized canoe ride down the canal from the guard lock to the Montgomery Mill. Her family owned the canoe. Her prettiness was a bonus.

Yet despite my youthful connections to the canal and its mills, I knew relatively little about the Windsor Locks Canal and its origins. I knew that my dad, who grew up across the street from the canal, used to spear shad from the canal bridge in front of what was once Haskell's Silk Mill and sell them for 10 cents apiece to Sisitsky's Market on Main Street. I knew that once a year, during "shut down," the canal was drained for two weeks so that it could be inspected and repaired. And I also knew, thanks to my Irish grandfather, Frederick Kervick, with whom I spent a great deal of time, that, "The Irish dug that canal."

I remember playing in my yard, and first hearing, then seeing a sweaty boy from the neighborhood *(I think it was Walter Martens)* running up the Elm Street hill shouting, "There's a boat in the canal, there's a boat in the canal." We all ran down the hill and, sure enough, there was a boat passing through the canal. I had never seen that before and never did again.

Twenty-two years ago, I was engaged by the State of Connecticut, Department of Environmental Protection, to complete a title search of the canal. More specifically, it was a title search of the canal towpath, which the State was in the process of leasing from the Dexter Corporation to make available for public recreation. It was a big project, and it took me over two months to complete. During the course of that search, I read a 1991 report prepared by historians Michael Raber and Patrick Malone. They completed the report at the request of the Connecticut

Department of Environmental Protection, Bureau of Parks and Forests. It was part of a study examining the feasibility of a Connecticut Canal Historic Exhibition Center. The report stated, "Relatively few of the nameless Irish workers, who lived near their work in temporary board houses or shanties, stayed after the canal opened in November 1829." I was incensed. I thought to myself, "They weren't nameless. Each one of them had a name. We just don't know it." That was the hook. My overblown sense of Irish indignation got the better of me. I determined then and there that I was going to find out everything I could about the canal and identify by name at least one of the Irish laborers who built it. That's been my pastime and passion for twenty-two years.

Like the Connecticut River, the intensity of that search experiences spring freshets and summer trickles, but always it flows forward. The search has brought me to archives, churches and cemeteries throughout the United States and Ireland. My wife and children have begrudgingly been to more canals, or former canal sites, than they care to remember, including a one-week cruise on a very slow Erie Canal barge.

It is not unusual that long lulls take place during the search. Sometimes, my professional and family responsibilities take precedence. Sometimes, the search is active but frustratingly unproductive. But on the opposite side of that same coin, there is no match for the excitement that is felt when an important discovery is made. When the archivist at the Archdiocese of Boston slowly walked into the reading room carrying Reverend R. D. Woodley's leather-bound, hand-written register of the baptisms he performed at the canal, I was physically shaking with anticipation. When I leafed through the pages and saw the words, "On the canal above Hartford," I knew that I was about to learn something that had been forgotten for nearly two centuries. When a box containing the personal files of the Chief Engineer was set on the table before me at the Cornell University Archives, and a manila folder labeled "Enfield Falls Canal" was among the other folders in the box, I knew that information believed to be lost to history was not in fact lost. A deeper understanding of how the canal came into being was about to be discovered. When I found in a tiny, rare bookstore, in an instance of sheer blind luck, the Locktender's ledger of expenditures from the canal's third year of operations, I celebrated my good fortune. These were truly exciting moments.

What follows are the results of my efforts. I hope you find them interesting. One spoiler - I succeeded in my goal of discovering the name of one laborer, and, as it turns out, a good many more. An alphabetical listing of these Irish laborers appears at the end of this book.

Finally, I wish to thank my wife, Michele, and our children Katie, Mollie, and Dan for their patience and support during the nearly quarter century of research that led to this book, and for their editorial assistance in the final push to publication.

A Note About the Name of the Canal

The purpose of the Windsor Locks Canal was to bypass the five-mile stretch of rapids on the Connecticut River known as the Upper Enfield Falls and Lower Enfield Falls. The project was first referred to as the "Canal at Enfield Falls" and upon completion the "Enfield Falls Canal." The upper guard lock and head gate are in the Town of Suffield, Connecticut. The lower set of three lift locks are in what at the time of construction was known as the Pine Meadow section of Windsor, Connecticut. Over time, as the manufacturing center at the lower end powered by water from the canal developed, the area became known simply as, "The Locks." As the area developed, a new post office was proposed to serve the needs of the growing industrial center. C.H. Dexter was appointed the first postmaster in 1843. At his suggestion, the name Windsor Locks was chosen to identify the new postal facility. The area was split off from Windsor and was separately incorporated as the Town of Windsor Locks in 1854. During this period, the name Enfield Falls Canal fell into disuse and the name Windsor Locks Canal became popular. The mild redundancy in the name Windsor Locks Canal seems not to have dissuaded anybody. The Windsor Locks Canal Company was incorporated in 1907. The Connecticut River Company, which had planned, constructed, and managed the canal was merged into the Windsor Locks Canal Company in 1926. There is no difference between the Enfield Falls Canal and the Windsor Locks Canal and the terms are interchangeable.

J. Christopher Kervick
Windsor Locks, CT
2025

CHAPTER ONE

The Push for Internal Improvements

The Connecticut General Assembly chartered the Connecticut River Company in 1824 to improve navigation on the Connecticut River. The purpose of the Connecticut River Company was to open commercial shipping to Barnet, Vermont some two hundred thirty-five miles upriver. Although those ambitious plans were never fully realized, the company succeeded in completing a canal bypassing the Enfield Falls. In addition to significantly enhancing river navigation, the canal became the backbone of a water-powered industrial center that remains profitable to this day. More significantly, the formation of the Connecticut River Company furthered Connecticut's position as a vanguard of the industrial revolution and demonstrated Connecticut's willingness to take bold action to retain its prominence. The ripple effect from this ambitious economic undertaking permanently expanded Connecticut's ethnic and religious demographics.

The story of Colonial America is largely that of its great cities. Boston, Philadelphia, New York, and Baltimore were all thriving economic, political, and cultural centers by the time of the American Revolution. The common denominator for each of these cities is that they were all accessible ports. The exchange of both goods and ideas was a vital component of city life. For the most part, these exchanges were conducted by sea.

Domestically, the cities were self-supporting enterprises. Agricultural products were locally grown and, with the exception of tobacco, intended for sale to nearby city dwellers. Manufacturing was conducted by craftsmen in small owner-operated shops within the cities. Manufactured goods were most commonly marketed on the local level.

Transportation by ship was always fraught with risks, but those risks paled in comparison to transportation by other means. Colonial roads were scarce and those that did exist were unreliable. Bridges were poorly constructed and were washed away with alarming frequency. Ferries dotted the landscape, but service was subject to interruption by spring freshets and winter ice.

As the colonial age gave way to the industrial revolution, the nation experienced a tremendous increase in the need for fast and more reliable means of transportation. As the population of the existing cities grew, and as more cities emerged, agriculture was pushed further inland where land and rich soil were plentiful. Delivering fresh produce to the population centers became more difficult. Similarly, as the population spread out over a greater area, distribution of manufactured goods became more complex. Dependable inland transportation was the key to this change, and without it, economic growth was impossible.

Connecticut was ahead of the game in the area of internal communication. In 1614, the Dutch explorer Adrian Block, aboard the *Onrust*, sailed the Connecticut River as far north as the Enfield Falls.[1] He recognized the river's potential as an aid to inland commerce. A Dutch fur trading post was soon established in Hartford.[2] The English settlers at Plymouth were equally enthusiastic. In 1633, Puritan settlers in Plymouth transplanted again to Windsor, where fertile meadows awaited the plow's edge.[3] Settlements at Wethersfield, Hartford, and Springfield followed.

On the Connecticut River, seagoing vessels transported manufactured and imported goods upstream as far as the Enfield Falls, which halted further progress. Those same vessels transported locally grown produce on the return trip. In 1638, William Pynchon of Springfield, established a warehouse on the east bank of the river at the base of the

---

[1] Henry R. Stiles, *The History of Ancient Windsor* (New York, C. B. Norton, 1859; republished Somersworth, NH, The New Hampshire Publishing Company, 1976), 17; Jabez H. Hayden, *Historical Sketches*, (Windsor Locks, The Windsor Locks Journal Publishing Company, 1900), p. 9.

[2] Stiles, *Ancient Windsor*, p. 18.

[3] Stiles, *Ancient Windsor*, 25; Hayden, *Historical Sketches*, p. 10

Headgate and guard lock of Windsor Locks Canal by Jack Fales.

falls.[4] In the warehouse, he stored goods awaiting overland transportation around the falls to Springfield. The warehouse also stored furs and agricultural products from the fertile Connecticut River Valley of Western Massachusetts, Vermont and New Hampshire awaiting shipment to the major population centers. The area became known as Warehouse Point and remains a village in the Town of East Windsor.

The natural boon to transportation that the Connecticut River provided was not without its obstacles. Overland transportation between Warehouse Point and Springfield was expensive and time-consuming. Shoals in the riverbed and low water during dry summers were obstacles to the deep draft sloops commonly used to transport goods on the river. Goods were frequently off-loaded to low draft flat boats, or scows, which were then poled over the rapids, but manpower costs were prohibitive.

---

[4] Hayden, *Historical Sketches*, p. 25.

As the eighteenth century closed, it was clear that further improvements were needed. Connecticut's natural advantages were no longer sufficient to serve the needs of a growing economy. Man would have to mold the earth to overcome the remaining geographic obstacles. Implementation of such large-scale public improvements required government involvement and more manpower than available at the time.

Early Public Improvement Projects

At the beginning of the industrial age, Connecticut used two different approaches to complete public improvements. One method, which was generally the earlier of the two, was for the State legislature or local governing body to compel the property owners along the route of the improvement to complete the construction. The owners were often taxed to fund subsequent maintenance. The second method was for the Connecticut legislature to charter a private for-profit corporation to build and operate a desired improvement. Investors funded the improvements by purchasing stock in the corporation anticipating returns from toll revenue. Some early examples of both methods follow.

The Great Meadow Drain

Proprietors of the meadow in Windsor north of the Farmington River between the Connecticut River and Palisado Avenue had difficulty farming the land because it was often too wet. The land became known as the "wet and drowned land."[5] In 1797, the owners petitioned Governor Oliver Wolcott requesting that a drain or sewer be permitted to drain off the excess water.

The Governor responded as follows:

To Mess. JABEZ HASKEL, DANIEL GILLET & EZRA HEYDEN all of Windsor in said STATE – GREETING

WHEREAS, upon the Memorial of James Hooker, Samuel Allen and others inhabitants of said town of Windsor showing that there is in said town on the East side of the highway on which they dwell a quantity of marshy low lands which are rendered unprofitable by the overflowing of waters being the whole

---

[5] Daniel Howard, *A New History of Old Windsor*, (Windsor Locks, The Journal Press, 1935), p. 315.

quantity of land contained in two former Commissions of Sewers and praying that Sewers might be appointed to drain the said land, THE GOVERNOR & COUNCIL on the 23rd day of May Adom. 1797 did appoint you the said JABEZ HASKEL, DANIEL GILLET & EZRA HEYDEN to be commissioners of Sewers to drain the land aforesaid.

I DO THEREFORE pursuant to the said appointment commissionate you the said JABEZ HASKEL, DANIEL GILLET & EZRA HEYDEN for that purpose, & to perform whatsoever is necessary & requisite thereto agreeably to the directions of the law entitled "An Act for appointing & directing commissioners of Sewers and Scavengers" being first sworn to act therein, & you are to conform yourselves to the provisions of the law to which your office hath relation.

GIVEN under my hand and seal in Hartford the 30th day of May Adom. 1797.[6]

To construct the drain, each proprietor was identified and given the responsibility for completing the section through his land. Proprietors subsequently elected three "scavengers" whose duty was to make sure that each owner cleared his section of drain at least once per year. The expenses of the scavengers were paid from the proceeds of a tax levied on each landowner benefiting from the drain.

Despite the dubious distinction of being appointed "Commissioner of Sewers and Scavengers," Jabez Haskel, Daniel Gillet, Ezra Heydon, and their successors apparently took their appointments seriously and were successful in opening and maintaining the drain until 1899.[7] The remains of Great Meadow Drain still serve its original purpose today.

Toll Roads

Connecticut's early highway system demonstrates an evolution from public works completed and maintained by compulsory labor to those operated by a corporate franchise. When the Native American trail running from Windsor to Hartford proved insufficient to handle increasing horse and cart traffic, the General Court at Hartford, in 1645, ordered:

---

[6] Ibid, p. 316.

[7] Ibid, p. 316

> Whereas there has been much dispute about the highway between Windsor and Hartford, which hath been lately used in coming through the meadow of Hartford with cart and horses, to the annoyance and prejudice of the inhabitants of Hartford that have lots in said meadow it is therefore thought meet and so ordered that the highway, as for carts, cattle and horses be stopped up, and that the said highway between the said Windsor and Hartford in the upland be well and passably amended, by Hartford as much as belongs to them, and Windsor as much as belongs to them, by each party in six weeks or two months on penalty of twenty shillings per week for each party that fails either in whole or in part.[8]

Although the inhabitants complied with the order, such forced labor must have been a great burden upon the citizens who had all they could do to grow, harvest, and store enough crops to feed their families. Although this system of road construction through forced labor continued through the eighteenth century, the roads it produced were "rock strewn, boggy, and craggy... a fright to travel over."[9]

In 1792, lacking sufficient funds to finance road construction on its own, the State of Connecticut granted its first franchise for the creation of a private toll road.[10] Although these toll roads were considered public, a private corporation was given the exclusive right to collect tolls from road users. Construction funds were raised from private investors. Toll revenue was collected to maintain the roads and, in theory, provide a return to investors. The typical toll road was about sixteen feet wide with drainage ditches on both sides.[11] The toll roads were called turnpikes due to the turnstiles at their tollgates. Labor on the toll roads was commonly provided by local farmers, hired by the day, often with teams of oxen. The labor was sporadic, as the farmers were only available during lulls in the farming cycle. It does not appear that the Turnpikes were ever profitable. Maintenance of the roads was often insufficient. Before Route 75 through Windsor Locks was renamed to honor Governor Ella T. Grasso, it was known as Turnpike Road, in reference to its inception as a toll

---

[8] Hayden, *Historical Sketches*, p. 42.

[9] Connecticut Department of Transportation, *DOT History*, https://portal.ct.gov/dot/general/history/chapter-1-dot-history?language=en_US, p. 1.

[10] *DOT History*, p. 2.

[11] Ibid.

The Old Ferry between Warehouse Point and Windsor Locks, CT.

road. By 1900, it was described as almost abandoned and overgrown with weeds.[12]

Whether completed by compulsory labor or a toll operator, Connecticut roads were insufficient to meet the growing transportation needs of the State. It was not until the introduction of heavy equipment, asphalt and automobiles that roads became reliable, and the preferred method of transportation.

Ferries and Bridges

Although the Connecticut River was a great aid to North-South transportation in the State, it was a great hindrance to East-West transportation. Additionally, tributaries such as the Farmington River posed obstacles that needed to be overcome.

In 1642, the General Court of Connecticut proclaimed that it would grant a license to any person willing to provide a ferry service across the Connecticut River in Windsor. The Court went so far as to propose a fee schedule allowing three pence for transport of a single passenger, two pence per person when the boat carried multiple passengers, and twelve pence for each horse. No citizen took the Court up on its

---
[12] Hayden, *Historical Sketches*, p. 46.

offer until 1649, when the Court contracted with John Bissell to:

> "keep and carefully attend the Ferry over the Great River at Windsor, for the full term of seven years from this day, and that he will provide a sufficient boat for the carrying over of horse and foot upon all occasions."[13]

The delay cost Bissell, as the toll for transport of persons remained as offered in 1642, but the toll permitted for transporting a horse dropped to eight pence.

Bissell's ferry, as it became known, with its relatively low start-up costs *(a boat)*, but its high operational costs *(one full-time operator)*, was organized as a private enterprise with a public function. The ferry was successful and remained in operation until 1917.

Other local ferries included a ferry across the Farmington River in Windsor from Ferry Lane on the north bank to the south bank east of Palisado Avenue; Wolcott's Ferry, which combined a ferry across the Farmington with a connecting ferry across the Connecticut River to Governor's Highway in what is now South Windsor; and a ferry from what is now Windsor Locks to Warehouse Point.[14] Each of these ferries was a private venture authorized by the State of Connecticut.

The East Windsor to Windsor Locks ferry rights were first granted by the General Assembly to James Chamberlain of East Windsor as follows:

> Resolved by the Assembly, that a ferry be allowed in the place prayed for and that the memorialist have an exclusive right to the same, provided he case out a road or highway from the common road in East Windsor through his land to the water's edge at the ferry place, and said ferry be under the same regulations as other ferries in said East Windsor." [15]

The "exclusive right" granted by the General Assembly proved valuable to the ferry operator. When Charles Jencks attempted to organize a competing service in 1835, the General Assembly refused to grant him a franchise.[16] Further, when a company was chartered to construct a bridge at the same point, as discussed below, the bridge company was forced to

---

[13] Howard, *A New History*, p. 299.

[14] Ibid.

[15] *Hartford Times*, July 9, 1909.

[16] Ibid.

pay the ferry company twenty thousand dollars for the franchise.[17]

While ferries served their purpose well for many years, they were always viewed as a next-best substitute for a good bridge. The problem was that construction methods and materials of the day were insufficient to build good bridges. From today's perspective, it is almost comical to consider the frequency with which the early bridges ended up washing down the river they were built to cross. Those that did not collapse often burned.

The earliest bridge in the vicinity of Windsor was built across the Farmington River in 1639-40. The bridge was constructed by the compulsory labor of local residents, including Reverend Ephraim Huit, who was reported to be so absorbed in his work that he had no time to visit with his fellow clergymen.[18] Reverend Huit's efforts provided only short-lived results, as the bridge had been washed away by 1645.[19] Other bridges crossing the Farmington were destroyed by flood "more than one" time between 1645 and early 1800's, and again in 1854.[20]

Connecticut granted a franchise to John Reynolds to construct a toll bridge across the Connecticut River from Suffield to Enfield in 1798. The bridge was not completed until 1808 and was gone some time before 1818. It was rebuilt as a covered bridge upon the original piers in 1832 and made it all the way to 1900 before it was swept away in February 1900, this time taking a gentleman by the name of Hosea Keach with it. After regaining consciousness, he punched through some loose boards to the outside, then surfed the bridge downriver until he was plucked out by workmen dangling a rope from the railroad bridge in Windsor Locks.[21] The bases of the stone piers of this fateful bridge still remain.

Hartford had its own problems keeping bridges in place. Jonathan Wolcott of Windham constructed the first bridge from Hartford to East Hartford. It opened in 1810 and was swept away by flood in 1818.[22] The bridge was replaced by a covered bridge the same year. That toll bridge served Hartford and East Hartford until it was destroyed by fire in 1895.

---

[17] Ibid.

[18] Hayden, *Historical Sketches*, 39.

[19] Ibid.

[20] Ibid.

[21] "Enfield Bridge is Gone," *Meriden Record-Journal* Feb 16, 1900.

[22] Hayden, *Historical Sketches*, p. 26.

The bridges crossing from Windsor Locks to East Windsor have had better staying power, predominantly because they came later when superior building materials and engineering methods were available. The first, a suspension bridge, was completed in 1886.[23] Again, the State chartered a private corporation to operate the toll bridge. Prior to the State purchasing the bridge and all other toll bridges over the Connecticut River in 1909, the Windsor Locks and Warehouse Point Bridge and Ferry Company paid nine percent per annum dividends to its shareholders.[24] The original bridge has been twice replaced and, to date, none of the three bridges has made a trip downriver.

Two Early River Channel Improvement Projects

In 1785, the following appeal was addressed to the Connecticut General Assembly:

> The Memorial of us the subscribers inhabitants of the First Society in East Windsor propriators & owners of land on the bank of Connecticut River about two miles in length humbly showeth—
> That the navigation of said river is of great importance to this State as thereby a large and beneficial trade is carryed on by sea, and into the large and extensive country northward the profits of which principally center in this State—That the navigation of said river is greatly impeded and obstructed by reason of barrs and shoals in it occasioned by sands washing into the brooks from the adjoyning roads and by the waters of said brooks carried into the beds of the river and abstruct and choak the same which occasions the waters of said river to wear away the banks of the river which consists of a fine loomy earth and the river thereby increases in width and decreases in debth so that in sommer seasons it is with great difficulty that even rafts of lumber and loaded boats pass by said town and your memorialists being fully convinced that said barrs and shoals may be removed by confining the river to narrourer limits and that the accretion to the adjoyning lands and their increased value would repay the expence to the propriators of said lands and are willing and desirous to make an attempt of this kind provided the design should meat with your honours patronage and your honours will compell the propriators of the adjoyning lands from the north side of Benja-

---

[23] Raber, Michael S. and Malone, Patrick M, Historical Documentation Connecticut Canal Historic Exhibition Center Feasibility Study and Master Plan, (prepared for Connecticut Department of Environmental Protection, Bureau of Parks and Forests, 1991), 27.

[24] *Hartford Times*, July 9, 1909

min Wolcotts lott to the north bounds of said society or so many of them as a commitee of disinterested & judicious men shall judg to be their just and reasonable part of the expence—Whereupon we the Subscribers humbly pray that your honours would appoint a judicious and disinterested committee and impower them at the expence of the propriators of the lands adjoyning to said river within the aforesaid bounds to errect such and so many wares & obstructions in said river within said limits as they shall judg nessary and convenient with power to tax the proprietors of said adjoyning lands or so many of them and in such proportion as said committe judg just and reasonable to defray the expense thereof with power to appoint collectors to gather said taxes.

Signed Erastus Wolcott, Aaron Bissell, Amasa Loomis, Elisha Bissell, Eli Moore [25]

This request was granted by the General Assembly, but it is not clear whether the improvements were completed. The remains of wing-dams designed to direct the flow of the river through the established channel exist to this day, but they are likely the result of later channel enhancement efforts. In this 1785 effort, the adjoining landowners volunteered to perform the labor associated with the improvement to obtain the anticipated benefits. As such, it would be more appropriate to characterize this public improvement as one completed with voluntary labor as opposed to compelled labor. Nevertheless, it quickly became apparent to the citizens and leaders of Connecticut that improvements of such magnitude could not continue to be completed and maintained with the part-time labor of local residents.

For channel improvements above Middletown to Hartford, the State chartered the Union Company in 1800. By 1806 the company had successfully added two feet to the controlling channel depth and began collecting tolls.[26]

The public improvement models of the colonial era and early industrial era were the necessary results of the labor shortages of the time. When the State ordered works to be completed by compulsory labor, it was because no other labor options existed. There was no established working class available to accept government-funded employment. Tax

---

[25] Howard, *A New History*, p. 75.

[26] DeLoss, Love, W. *The Navigation of the Connecticut River*, (American Antiquarian Society, April 1903), 899.

revenues were insufficient to fund such projects even if the manpower was available. Government chartered monopolies provided a partial solution to the difficulty of completing public improvements. But these enterprises also frequently suffered from low revenues and a shortage of workers. As the need for public improvements continued to grow, and the scale of improvements increased, the labor shortage became more pronounced and creative solutions were required.

CHAPTER TWO

Muscle From Abroad

As the push for internal improvements was intensifying in the United States, social and political changes were transpiring in Ireland. These changes would set the stage for a wave of Irish immigration to America. The most familiar and largest wave of Irish displacement occurred during the "potato famine" years of 1845 to 1855. Yet a generation earlier, America experienced its first significant wave of Irish immigration, and it is this first wave that produced the laborers needed to sustain America's canal building boom.

The Evolution of Irish Poverty

John O'Beirne Ranelagh's outstanding book, *A Short History of Ireland*, provides a succinct review of Ireland's history.[1] That book, and the several additional sources cited herein, were relied upon heavily in the construction of the following paragraphs.

After St. Patrick's conversion of Ireland to Christianity near the end of the Fifth century, Irish culture and scholarship flourished. During the Dark Ages, while ravaging tribes victimized the remainder of Europe, Ireland was left unscathed and enjoyed a period of enlightenment. Many of the foremost European scholars and philosophers took refuge in Ireland and contributed to the social and intellectual awakening.[2] However, the year 795 marked the beginning of nearly two centuries of Viking

---

[1] Ranelagh, John O'Beirne, *A Short History of Ireland* (Cambridge: Cambridge University Press, 1983).

[2] Ranelagh, *A Short History of Ireland*, 26; Thomas Cahill, *How the Irish Saved Civilization* (New York: Anchor Books, 1995).

raids and occupation.[3] By the time Brian Boru defeated the Danes and was acknowledged as the first High King of Ireland following the Battle of Clontarf in 1014, the Danish influence in the coastal cities of Dublin, Wexford and Waterford was well-established. The Danes had accepted Christianity, and the coastal cities became bustling trading and commercial centers. This area became know as the Irish Pale. The remainder of Ireland, or "Beyond the Pale" as the expression developed, continued to be little influenced by a changing world.[4]

Gaelic culture and customs beyond the pale differed significantly from that of feudal England and Europe. Irish social structure and property rights were determined by the Brehon Laws. Unlike English feudal law, where all land was owned by the king, who granted fiefs and land titles to nobles who met with his favor, and who in turn granted parcels of land to tenants and subtenants, under Brehon law, title was granted not to individuals but to families or clans. All adult males with a common great grandfather shared title.[5]

Similarly, Christianity beyond the pale was little influenced by the Roman Church. The Brehon laws influenced the Irish clergy. Monasteries were closely associated with the ruling families and the abbots passed their offices from generation to generation. This process naturally resulted in a clergy more concerned with their benefactors than with commoners.[6]

The widening gap between English and Irish culture and the growing independence of Irish Catholicism were causes for concern both in England and in Rome. Those concerns were acted upon when the reign of Pope Adrien IV, the first and only English Pope, coincided with the reign of the Norman-English King Henry II. Adrian IV asserted papal supremacy over Ireland and authorized Henry II to enter Ireland to enforce it. So began the Norman-English invasion of Ireland which culminated in the late 12th century.[7]

---

[3] Ranelagh, p. 27; Cahill, p.211.

[4] Ranelagh, p. 31.

[5] Catherine Duggan, *The Lost Laws of Ireland* (Dublin, Glasnevin Publishing, 2013).

[6] Ranelagh, p. 31.

[7] Ibid.

The Normans sought to replace the Brehon laws regarding land ownership with English Common Laws. They built castles and walled cities to protect their lands and their claims upon them from the generally hostile and resentful Irish.[8] The Normans enacted penal laws, including a ban on intermarriage and a ban on speaking Irish, to prevent the Norman-Irish from assimilating into Irish culture.[9] Despite their best efforts, by the end of the fourteenth century, the Normans in Ireland had become more Irish than English. English influence beyond the Irish Pale was almost non-existent.[10]

Ireland moved closer to independence over the ensuing years until the reign of the Tudor King Henry VIII. Henry VIII believed that Irish rents had long been a neglected source of revenue for the crown.[11] He enacted Surrender and Re-grant legislation, under which all Irish land titles reverted to him, and he re-granted it only to those loyal to him. These land policies were often implemented through the establishment of plantation schemes where thousands of Scottish and English settlers were awarded lands, especially in Northern Ireland.[12] This marked the end of influence of the Brehon laws upon Irish land ownership. Henry VIII's split with Rome and his creation of the Anglican Church further inflamed Irish passion. The Catholic Mass was banned in 1549. Catholics were fined for failing to attend Church of Ireland services.[13] What had previously been primarily a dispute over land, took on much more of a religious character, Catholic versus Protestant, as a result of these changes.[14]

The divisions culminated in revolution in 1641, marked by the massacre of 12,000 Scottish and English planters in Ulster County.[15] The Confederation of Kilkenny, a combination of Catholic clergy, na-

---

[8] Ibid.

[9] Ranelagh, p. 39; Duggan, p. 120.

[10] Ranelagh, pp. 41-42.

[11] Ibid., p. 46.

[12] Ibid., p. 49.

[13] Ibid., p. 50.

[14] Ibid., p. 56.

[15] Ibid., p. 61.

tive Irish leaders and Old English transplants to Ireland, was formed in 1642. For a brief time, it looked like the Confederation could secure a peaceful resolution to the revolt, by maintaining allegiance to the King in exchange for relief from oppression. However, the English parliamentary revolution, the execution of King Charles I, and the rise to power of Oliver Cromwell brought an end to those hopes. Cromwell made the submission of Ireland a top priority. He arrived in Dublin in 1649 with an Army of 12,000 men and began a campaign of conquest and ruthless slaughter. Cromwell dispossessed all Catholic landowners east of the Shannon River. Their land was redistributed among Cromwellian soldiers and English investors.[16]

To keep the Irish subdued, the Irish Parliament, composed entirely of British landholders, enacted the Penal Laws. The first was a ban on the possession of any weapon. To quell the influence of religious leaders, all Catholic bishops were banished from Ireland and all remaining priests were required to swear allegiance to the crown and to be registered, no more than one to each parish. To assure Protestant dominance over property, Catholics were banned from inheriting land from Protestants, Catholics could not purchase land or lease land for more than thirty years, and existing Catholic landowners were required to will their land equally to each son, unless a son converted to the Church of Ireland, in which case that son would inherit the entirety. The Penal Laws also denied to Catholics the right to vote, to receive a formal education or to enter a profession.

The enactment of the Penal Laws, while perhaps not the source of Irish poverty, certainly guaranteed the perpetuation of poverty. As the eighteenth century dawned, most rural Irish lived in small communities known as clachans. These were clusters of stone or mud huts, upon communally owned parcels of land. The residents were subsistence farmers. The inheritance aspect of the Penal Laws resulted in the subdivision of these lands with each generation and the inability to expand the holdings. Increasingly, rural farmers were left with insufficient land to support their own families. As the century progressed, a world market for Irish goods developed. In particular, the demand for Irish grain and linen increased dramatically. While the increased demand improved the Irish economy generally, it did little to enhance the lives of the rural Irish.

---

[16] Ibid., pp. 62 - 65.

Contemporary illustration of dire conditions Irish families endured due to the English Penal Laws.

Protestant landowners rapidly expanded their farms to meet the growing market demands. The Penal Laws facilitated these acquisitions. The clachans began to give way to a landless peasantry that depended on the Protestant farmers for employment and shelter. Such an existence was tenuous, as the Protestant farmers controlled both their wages and their rents, and could therefore keep the peasants on the edge of existence. As the eighteenth century came to a close, the rural Irish were between a rock and a hard place. The remaining clachans were overpopulated and could no longer provide even bare subsistence, and the tenant farmers faced eviction with no place of refuge.

Faced with such desperation, it was not long before the oppressed began to fight back. Secret societies emerged and violence became commonplace. Evicting landlords were a frequent target. In Northern Ireland, working class Presbyterians, the descendants of transplanted Scotts, formed the United Irishmen. Desperate Catholics formed the Defenders. Both movements were heavily influenced by the ideals of the American and French revolutions. Revolutionary France declared war on Britain in 1793 and vowed to aid any nation intent on overthrowing oppressive rulers. Anticipating the spread of revolution to Ireland and the aid of French troops, the United Irishmen and the Defenders joined forces in 1796. The Anglican aristocracy in Ireland reacted aggressively to this threat by infiltrating the organization and arresting most of its leaders. Ireland was at the boiling point.

Canal Laborers Michael Costello and Lawrence McMahon

It was into the perilous state of affairs that the future Windsor Locks Canal workers were born. Included among them were Michael Costello and Lawrence McMahon. Costello was born in 1793 and Lawrence McMahon in 1800.

Bust of Stalker Wallace

Michael Costello

Michael Costello's parents were from the Catholic Parish of Kilfinane, in County Limerick. Around the time of St. Patrick's conversion of Ireland to Christianity in the later half of the fifth century, St. Finnian resided in a small hut on the banks of the Lubagh River. He led a monastic life filled with daily prayer and meditation. The peaceful countryside provided an ideal setting for his prayerful life, and it was not long before curious local farmers began to emulate his lifestyle and beliefs. After St. Finnian left the area, the local residents erected a church in his honor, naming it Cill Finnian, or the church of Finnian. The local area became known as Kilfinane.

But the Kilfinane of the late eighteenth century was no longer the peaceful and idyllic setting of St. Finnian's meditations. When Michael Costello was a boy of just five years of age, Kilfinane was rocked by the Staker Wallis incident. The Catholic Parish of Kilfinane included the Civil Parish of Kilfinane and portions of the civil parishes of Particles and Ardpatrick. Twenty-four townlands within this area were owned by Charles Silver Oliver, the grandson of Captain Robert Oliver, who had acquired the land as a reward for his service in Cromwell's army.

As a landlord, Charles Silver Oliver, was a natural target of the United Irishmen, and he learned that a tenant farmer by the name of Patrick Wallis was conspiring to assassinate him. Wallis, a man of sixty-five years of age, was the leader of the local company of United Irishmen. For his protection, Oliver had pressed into service the young sons of many local farmers who were adept at riding to form a yeoman's cavalry. Most were reluctant to serve but had no say in the matter. On a foggy Sunday morning in March of 1798, they set out from Castle Oliver to Kilfinane to capture Patrick Wallis. When Wallis learned of their approach, he set out on horseback for the safety of the United Irishman hideaway in the mountains. A chase ensued. A young man named

Michael Walsh gained on Wallis but, reluctant to capture him, he drove his horse into a bog and was nearly drowned in the muck. Eventually a young rider named Roger Sheedy caught up with and captured Wallis and he was taken to Kilfinane jail.

In jail, Wallis was offered his freedom in exchange for revealing the names of each of his co-conspirators. He remained silent. Oliver ordered Wallis stripped and tied to a cart and whipped as he was led through the streets of Kilfinane, but again he remained silent. Wallis was then brought to a cattle fair in nearby Ballinvreena where he was publicly whipped. When he remained steadfast in his silence, Wallis was returned to Kilfinane and hanged. Oliver ordered that Wallis be beheaded and his head was set high on a stake above the market house in the town square. Wallis was thenceforth revered and remembered by his macabre nickname, "Staker Wallis". The haunting air, "The Lament of Staker Wallis," has become a favorite tune of Irish pipers and violinists.

For his muddy detour, Michael Walsh was highly regarded by the locals. To the contrary, Roger Sheedy paid dearly for his efforts. *The Freeman's Journal* of April 3, 1800 reports:

LIMERICK, MARCH 29

> On Thursday night last, a number of rebels murdered two men of the name of Sheedy, father and son, near Kilfinan, in this county; they broke open the house, and shot the old man instantly, but the son contrived to get out; they pursued and overtook him, on which the murderers told him that he ran very well after Staker, (a rebel who the young man apprehended, and was executed two years ago at Castle Oliver) but he must now run a different course, on which they put an end to his existence. The unfortunate victims were remarkable for loyalty; the son, a very fine lad, and a member of Captain Oliver's corps of yeomanry.

The Staker Wallis incident must have had a profound effect on the young Michael Costello, as well as all the young men and women of Kilfinane who would come of age over the next twenty-five years. The Wallis incident is but one example of the violence and desperation that had spread throughout Ireland, especially the rural and agrarian southwest.

Lawrence McMahon

Although the actual townland of Lawrence McMahon's birth remains undiscovered, census records indicate that he was born in Ireland around

1800 and church records show that he was married to Johanna McCarthy at St. Munchin's Church in Limerick City, Ireland on May 21, 1828. He is most likely from one of the agrarian communities in northern Limerick County or southern Clare County that encircle the city and were part of the Catholic Diocese of Limerick. Like the agrarian communities of southern Limerick, these communities were also in civil distress at the time of Lawrence McMahon's birth and during his boyhood.

Michael Costello and Lawrence McMahon were just two of the thousands of Irish children whose formative years were defined by oppression, poverty, violence and desperation. As these children matured, the conditions of their youth produced a hopeless Irish generation. For them, there was no future in Ireland.

CHAPTER THREE

The Dawn of the Canal Era

Canals Abroad

Traders in China and Egypt are known to have used locks and channels to facilitate the transportation of goods. These technologies eventually spread to Europe with the opening by Marco Polo and others of trade routes to the Orient. Great Britain and Ireland developed an extensive canal network to facilitate internal trade.[1] Due to the dearth of experienced civil engineers in America, New York Governor DeWitt Clinton, the chief proponent of the Erie Canal, sent a promising young engineer named Canvass White to England in 1817 for one year to study that country's extensive canal system. White applied what he learned in England to assume a major role in Erie Canal design and construction.[2] He would also later be engaged as the Chief Engineer of the Enfield Falls Canal.

Early U.S. Canals

The birth of the canal movement in America occurred on the Connecticut River just a few miles north of the Enfield Rapids. The Enfield Falls are one of six significant falls between the mouth of the Connecticut River at Old Saybrook and Northern Vermont. The others are the South Hadley Falls in Massachusetts, and in Vermont, Millers Falls, Bellows

---

[1] Drago, Harry Sinclair, *Canal Days in America,* (New York Clarkson N. Potter, 1972), p. 5.

[2] Bernstein, Peter L., *Wedding of the Water*s, (New York, W. W. Norton & Co., 2005), p. 192.

Falls, Water Quechy Falls and Orcott's Falls. Unlike the Enfield Rapids' gradual descent of thirty feet over five miles, each of these upper falls consist of deeper drops over shorter distances. Despite the inefficiency of the process, the falls at Enfield could be overcome under most conditions by a crew of stout Warehouse Point men poling scows or flat-boats against the rapids. The upriver falls represented absolute barriers to navigation. All goods had to be offloaded, transported overland by wagon to a point above the falls, and then reloaded onto vessels for further upriver transportation.

At South Hadley, where the river drops fifty-three feet in a little over two miles, several local business and political leaders combined to form a private company incorporated by the Massachusetts legislature as the Proprietors of Locks and Canals on the Connecticut River. The company was incorporated in 1792 and the South Hadley Canal was opened for business in 1794.[3]

The design of the South Hadley Canal had an unusual feature. Rather than using a system consisting solely of locks and gates to lift boats over the falls, the original canal utilized an ingenious device known as an "inclined plane" to lift and lower boats. The masonry stone plane was fifty-three feet in height at the high end, and declined at thirteen and one-half degrees over a length of two hundred and thirty feet to its base. A wheeled cart supporting a framework sufficient to secure and carry a ten-ton flat-boat with its thirty tons of cargo was attached to two chains. The chains were likewise attached to a winch powered by two sixty-foot diameter overshot water-wheels. Upon entering the lower gate, a flat-boat was floated above the cart, secured to the framework, and the cart and boat raised along the incline by letting water onto the giant water-wheels through a sluice. The wheels of the cart were sized so that while upon the plane, the cart and the piggy-backed flat-boat remained level.[4]

Variations of this design were subsequently incorporated into the plans of the Morris Canal in New Jersey. The device was particularly effective in canals where the type of vessel in use was of uniform size and shape, such as the coal barges of the Morris Canal. In a river improvement project, where considerable variety existed among styles and

---

[3] Drago, p. 14.

[4] Ibid., p. 15.

dimensions of boats involved in transportation, the design was less effective. Nevertheless, the design of the South Hadley inclined plane was so creative that the canal quickly became a popular tourist attraction, as throngs of people came to witness its operation. Ultimately, the novelty of the design could not overcome its limitations, and traditional locks to replace the inclined plane were completed in 1805.[5] The South Hadley project proved to be a long-term success. For the better part of half a century, the South Hadley canal facilitated the efficient transportation of goods and passengers upon the Connecticut River.

Subsequent canal projects were soon undertaken to bypass the remaining upriver falls. Quechy Falls was bypassed by a short canal and two locks completed in 1800. Miller's Falls was tamed by a three-mile canal utilizing eight locks completed in 1801. The three-quarter mile Bellows Falls canal with seven locks was completed in 1802 and Orcotts Falls with a short canal and five locks opened in 1810.[6]

While the Connecticut River projects are generally not viewed as true overland canals but river improvements, elsewhere in New England, a more ambitious canal project was underway.

At the close of the Revolutionary war, and during the period of economic recovery thereafter, the thriving port of Boston still found itself wrestling with the problem of internal communication. The same forces that brought settlers from Plymouth around Rhode Island to settlements in the Connecticut River Valley were still at work, one hundred years later, in Boston. The difficulty and expense of transporting goods overland to the New England heartland continued to frustrate the wealthy merchants and traders of Boston. The improvements on the Connecticut River were beginning to provide some relief in reaching Western Massachusetts, Vermont, and Western New Hampshire. Yet, there was still no reliable water route north from Boston into Southern and Central New Hampshire.

In 1792, James Sullivan, a renowned and wealthy Boston lawyer, who recently had become the attorney general of Massachusetts, teamed with Laommi Baldwin, a self-made military engineer who had risen to

---

[5] Ibid.

[6] *Report of the President and Directors of the Connecticut River Company: With the Report of H. Hutchinson, Esq. Laid Before the Stockholders at their Annual Meeting, January 3rd, 1826.*

the rank of colonel during the Revolutionary War. The pair championed the cause of an overland canal from Boston to the Merrimack River Valley.[7] Their petition to the legislature was accepted, and on June 22, 1893, Governor John Hancock signed into law the act incorporating the Middlesex Canal.[8] The act granted the corporation the right of eminent domain to acquire property along the proposed canal route and the right to charge canal users a toll of 5.5 cents per ton for goods shipped on the canal. The act required the project to be completed within ten years.

Because there was no established civil engineering profession in the United States at the time, the company had difficulty finding a person or firm qualified to survey a possible route. Accordingly, Laommi Baldwin and a team of inexperienced surveyors conducted an initial survey, which contained so many errors, and errors of such scale, that the survey was almost worthless. Baldwin went to Philadelphia and Washington seeking a qualified surveyor and construction superintendent. In Philadelphia, he met William Weston, an Englishman who was an experienced canal engineer, and who was engaged in several projects in that area. Weston came to Massachusetts and spent nine days surveying two possible routes. Weston favored the eastern route, which passed through the towns of Stoneham and Reading, but the inhabitants of those towns put up stiff resistance to the commercialization of their rural towns. Ultimately, the western route passing through the towns of Woburn and Wilmington was chosen.

The property acquisition phase went smoothly. The Company needed to resort to its power of eminent domain for only sixteen of the one hundred forty-two parcels acquired. Despite his difficulties with the initial survey, Baldwin was chosen to serve as Superintendent of Construction.[9] The decision was a wise one, for what Baldwin lacked in experience, he made up for in creativity and determination.

The plans for the 27.5-mile canal called for a trench, twenty feet wide at the bottom and thirty feet wide at the top, a ten-foot wide towpath, twenty locks, eight aqueducts, and fifty bridges. Baldwin incorpo-

---

[7] Drago, pp. 9-14; Ronald E. Shaw, *Canals for a Nation*, (University Press of Kentucky, 1990), p. 11; *Middlesex Canal Corporation Records*, University of Massachusetts, Center for Lowell History.

[8] *Middlesex Canal Corporation Records*, University of Massachusetts, Center for Lowell History.

[9] Shaw, p. 13.

rated, with some modifications, canal building methods and materials first introduced in England. Baldwin developed an American version of hydraulic cement, using West Indian volcanic ash as a binding ingredient, which provided the strength necessary to hold together the large stones used in constructing the lock chambers. Baldwin also developed a simpler method of "puddling", in which one or more layers of clay are used to prevent water from filtering out of the canal bed. Labor for construction was solicited from the property owners themselves, whose land the Canal crossed.[10] The plows and teams utilized on their farms were re-directed toward cutting the trench. The reliance on a local labor force resulted in some delays, as the first responsibility of the farmers remained with their fields.

Despite Baldwin's creativity, under-capitalization of the project put him at constant odds with the Board of Directors. Frequently, sound engineering considerations were ignored in favor of more cost-effective means of construction. In particular, Baldwin was forced to construct many of the locks with yellow pine, rather than the more stable, but more expensive, stone.[11] Despite the engineering challenges, and the financial battles, Baldwin pressed forward, all the while with the full support and confidence of James Sullivan.

By November 7, 1797, only the first six miles of Canal were open. By the end of 1798, only ten miles were open. Tension between Baldwin and the Board of Directors increased, and the Board openly solicited a replacement for him. Baldwin persevered, and on December 31, 1803, less than twenty-four hours before the expiration of his ten-year deadline, the entire length of the canal was completed.[12]

During its first five years of operation, many of the engineering shortcuts forced upon Baldwin came back to haunt the corporation. Frequent need for repairs hampered efficiency and made for an unreliable cash flow. In 1808, both Laommi Baldwin and James Sullivan died. Although these visionaries would never live to see the full extent of their dreams realized, their deaths did not signal the demise of the Middlesex Canal. To the contrary, following their deaths, Sullivan's son, John

---

[10] Drago, pp. 9 - 12.

[11] Shaw, p. 13

[12] Ibid., p. 13

Langdon Sullivan, assumed the position of Superintendent of the Canal, and his energetic and efficient management of the operation quickly reversed its fortunes. Needed repairs were made, tolls were collected, and by 1819, the Company paid its first dividend to shareholders.

John L. Sullivan *(not to be confused with the bare-fisted pugilist by the same name who would dominate the boxing world by the end of the century)* provided the energy and leadership needed to successfully realize the vision of his father and Baldwin. But the younger Sullivan was a visionary in his own right, and he expanded the enterprise. John L. Sullivan championed efforts to make river improvements on the Merrimack River. By 1814, a fully navigable route had been established from Boston to Concord, New Hampshire. Sullivan also began to experiment with steam engines during this period. Although the wake from his experimental steam-propelled barges often washed out the earthen banks of the canal, their introduction foreshadowed the combination of technologies that would make the Windsor Locks Canal a reality.[13]

Early Efforts to Circumvent the Enfield Falls

John Reynolds, born June 8, 1738, was the first son of Reverend Peter Reynolds, a minister of the Enfield Congregational Church from 1725 to 1768. Reynolds was a Selectman in the Town of Enfield and appears to have been well acquainted with the issue of establishing a reliable river crossing from Suffield to Enfield. As early as 1778, the Town of Enfield requested that the General Assembly transfer ownership to the Town of a ferry operated by Isaac Kibbe crossing the river just north of the mouth of Freshwater Brook.[14] Again in 1785, a committee was formed to negotiate directly with Isaac Kibbe for the purchase of the ferry.[15] Kibbe wouldn't budge. John Reynolds became a member of the third committee formed by the Town to acquire the ferry rights in 1796,[16] a fourth in

---

[13] Ibid., 13

[14] Enfield Town Meeting Minutes, December 7, 1778.

[15] Enfield Town Meeting Minutes, April 11, 1785.

[16] Enfield Town Meeting Minutes, April 11, 1796.

Depiction of flatboats on the Connecticut River north of Hartford.

1797,[17] and a fifth committee formed in 1798.[18] Kibbe held firm, and the General Assembly made no effort to divest him of his franchise. Frustrated, Reynolds came up with an alternative plan.

In October of 1798, Reynolds and several other residents of the Towns of Suffield and Enfield, petitioned the Connecticut General Assembly for permission to erect a toll bridge across the Connecticut River at Mad Tom Bar. Local mariners were critical of the plan, claiming that such a bridge would further impede the already difficult navigation of the Enfield Rapids. Reynolds maintained that any disruption to navigation created by the bridge could be overcome by the construction of two locks on the East bank of the river to bypass Mad Tom Bar and Surf Bar. Surf Bar, about one-quarter of a mile south of the present Route 190 Bridge, and Mad Tom Bar, about one-half mile south of the 190 bridge, are each formidable sections of the upper Enfield Rapids.[19]

The General Assembly granted the petition and incorporated The Company for Erecting and Supporting a Toll Bridge with Locks from Enfield to Suffield. The franchise gave the Company the authority

---

[17] Enfield Town Meeting Minutes, April 10, 1797.

[18] Enfield Town Meeting Minutes, April 9, 1798.

[19] *Resolve Incorporating the Company for Erecting a Toll Bridge, with Locks, from Enfield to Suffield. Private Laws of Connecticut*, General Asser, Vol. I,p. 249; Harte, *Connecticut Canals, The Fifty-Fourth Annual Report of the Society of Professional Engineers*, 193 8.

to construct the bridge, to collect a toll from persons using the bridge and to collect a toll from vessels utilizing the locks at the rate of thirty-four cents per ton. The franchise was to last one hundred years or until the Company had recovered its construction costs plus twelve percent interest.[20]

The General Assembly specified that the bridge must be no less than twenty-five feet wide and that the locks no less than thirty feet wide. What proved to be the most onerous specification by the General Assembly was the requirement that the project be completed within six years. Failure to meet the construction deadline meant forfeiture of the franchise.

The most valuable aspect of the franchise was the monopoly on bridge traffic north of Hartford. The General Assembly ordered that, provided the bridge and locks were maintained in good repair, "no person or persons shall have liberty to erect another bridge anywhere between the North line of Enfield and the South line of Windsor." The Locks, while also a monopoly, were perceived as less valuable, perhaps due to anticipated high construction or operational costs. In fact, several years later, when the Connecticut Supreme Court was interpreting the legislature's intent in requiring the company to complete the Locks, it determined that the General Assembly had not bestowed a right on the company, but imposed a burden.[21]

The task of construction, or more likely the task of funding construction, proved to be much more difficult than anticipated. In May 1805, the company petitioned for, and was granted, an extension of the completion date to June 1, 1808.[22] All other rights and obligations were ordered to remain the same. The company, again in May 1806, requested a further extension and the request was granted, extending the completion date to 1809.[23]

---

[20] *Resolve Incorporating the Company for Erecting a Toll Bridge, with Locks, from Enfield to Suffield. Private Laws of Connecticut*, General Asser, Vol. I, p. 249

[21] The Company for Erecting and Supporting a Toll Bridge with Locks from Enfield to Suffield v. The Connecticut River Company, 7 Conn. p. 28 (1828).

[22] *Resolve Incorporating the Company for Erecting a Toll Bridge, with Locks, from Enfield to Suffield. Private Laws of Connecticut, Private Laws of Connecticut*, Asser, Vol. I, p. 251.

23 Ibid.

Postcard of second Enfield Bridge by W.W. Chesnut. The bridge was built on piers of the original bridge. It was destroyed by flood in 1898.

In May 1808, the company sought additional concessions. By that time, the work on the bridge was underway, but the company now insisted that a channel in the water along the eastern bank of the river would improve navigation better than the proposed pair of locks. The company requested that the original grant be modified to incorporate this engineering change. The General Assembly relented and authorized the company to "make a shore channel, by excavating the bed of the river, or in such other way as they shall deem proper, over the aforesaid rapids, in lieu of the locks provided in said grant." The company was authorized to collect the same toll for passage through the shore channel as they would have collected for passage through the locks.[24]

The members of the General Assembly must have begun to wonder if the requested change was driven more by economic considerations than it was by engineering considerations. Their suspicions thus aroused, the General Assembly, with this last amendment, decided that it needed to look deeper into the engineering issues. The Assembly appointed Oliver Mather, Josiah Bissell, and Shubael Griswold as a committee to examine and inspect the bridge in progress, the plans for the locks first proposed, and the substituted plan for a shore channel. The committee

---

[24] *Resolve Incorporating the Company for Erecting a Toll Bridge, with Locks, from Enfield to Suffield. Private Laws of Connecticut, Private Laws of Connecticu*t, Asser, Vol. I, p. 251.

was ordered to determine whether the shore channel was sufficient to answer the intended purpose of improving navigation.

In October 1808, the company submitted a further petition. It reported that the bridge was nearly complete and would be ready to be opened in November of that year. As to the locks or shore channel, it represented that it had met with unforeseen difficulties and that, upon further examination, it had determined that a shore channel could not be completed, except at enormous expense, and its utility was doubtful. Despite the apparent disingenuousness of the claim, the company now asserted that locks, after all, would be more appropriate than a shore channel, but that in order to fill the locks with water, it would need to construct either a full dam across the river or substantial wing dams. Since a dam was not part of the original grant, the company maintained that it was without authority to continue. The company invited the General Assembly to appoint a second committee to examine the entire issue and make recommendations on the most expeditious manner of locking the falls.

By successfully muddying the waters as to the appropriate manner of improving *(or at least not further impeding)* navigation, the Reynolds group had avoided the need to make any further investment in the enterprise other than the costs of constructing the bridge. Boldly, the company sought permission to open the bridge and begin collecting the tolls.

The General Assembly must have realized it had been tricked, but there was also no point in preventing the opening of the bridge. The General Assembly authorized the company to open the bridge and begin collecting tolls once construction was complete in November 1808. This time, however, the General Assembly, once fooled, determined that it would not be fooled again. The Assembly added a heavy-hitter to the committee to decisively determine the best way to navigate Mad Tom and Surf Bar, whether that involved locks, a shore channel, some combination of both, or some entirely new method. Eli Whitney, the legendary New Haven inventor, was added to the committee, at the expense of the company, to examine the issue and report to the General Assembly in the spring. The committee was not ready to report its findings by the following May, but in October 1809 the committee made a surprising report to the General Assembly. Whitney and his fellow committee members reported that one lock at Mad Tom Bar was sufficient provided a full dam across the entire river at the same point was constructed. The dam would need to be of such height that it would raise the level of the river north to the point that Surf Bar became a non-factor. The committee further suggested that the entire enterprise would have much greater public utility if it could be combined with a plan to bypass the impediments posed by the lower Enfield Rapids. The committee suggested that any further efforts to improve navigation around Surf Bar and Mad Tom Bar be postponed until a corporation willing to undertake the entire project emerged.

This report must have delighted the Reynolds group as much as it dismayed the General Assembly. Nevertheless, the esteemed Eli Whitney had spoken, and the General Assembly was not prepared to challenge him. Reluctantly, but with no viable alternative, the General Assembly resolved in October 1809, that "the building of locks upon said falls, called Mad Tom Bar and Surf Bar, by said Company, be suspended, and the Company discharged from the obligation to build the same, until further order of this Assembly." It was further resolved, "that the petitioners be authorized to take the toll prescribed by their charter, of passengers crossing said bridge, the same being finished, and having been approved by the committee appointed by this Assembly for that purpose."[25]

The Enfield Bridge Company's victory was complete. It had out-maneuvered the General Assembly and secured the lucrative bridge monopoly with a minimal investment. Despite the fact that the bridge added a further impediment to the already dangerous navigational hazards, the company relieved itself of any responsibility for aiding navigation on the river.

In the end, the mighty Connecticut River had the last say, as the company's bridge was soon destroyed by flood, and only the stone piers remained by 1818. After several transfers of the franchise, the bridge was rebuilt as a covered bridge upon the original piers in 1832, and it struggled through several additional ownership changes and poor maintenance until it closed for good in 1896. In February 1900, the bridge came to a dramatic end when a flood carried the main section off its

---

[25] *Resolve Incorporating the Company for Erecting a Toll Bridge, with Locks, from Enfield to Suffield. Private Laws of Connecticut, Private Laws of Connecticut*, Asser, Vol. I, p. 252; The Company for Erecting and Supporting a Toll Bridge with Locks from Enfield to Suffield v. The Connecticut River Company, 7 Conn. p. 28 (1828); *An Act to Incorporate John L. Sullivan and His Associates for the Purpose of Construction Locks, Dams, and Channels at Enfield Falls, Private Laws of Connecticut*, Asser, Vol. 1, p. 507.

piers.²⁶ The bases of the stone piers of this fateful bridge still remain, and still pose a hazard to those navigating the Enfield Rapids. Following the General Assembly's suspension of the company's obligation to construct locks or some other aid to navigation, no further plans emerged until 1818. In October of that year, John L. Sullivan, who was rising to prominence as superintendent of the Middlesex Canal in Massachusetts and who was the driving force behind the Merrimack River improvements, directed his attention to improving the Connecticut River. Sullivan envisioned a shipping network from Boston utilizing a canal through the neck of Cape Cod, along Long Island Sound, and up the Connecticut River to reach Western Massachusetts and Vermont. In an open letter to the citizens of Hartford, Sullivan promised:

> Thus your city will be made the entreput of the trade of Boston, with the counties of Hampshire, Hampden, and Berkshire, at much less expense as direct land carriage now costs – trade, commerce and agriculture will, by increased facilities be improved; and Hartford become the mart of much more produce for exportation than at present.²⁷

Upon Sullivan's petition, the Connecticut General Assembly incorporated The Proprietors of the Enfield Locks and Channels.²⁸ This second effort to construct a functional bypass of the Enfield Falls was doomed from the start. The General Assembly so intertwined the interests of Sullivan with the interests of The Enfield Bridge Company, that no amount of enthusiasm and creativity on the part of Sullivan could make the enterprise work. Although John Reynolds had been dead since 1812, his successors had inherited his characteristic obstructionism, bumbling and shortsightedness.

Rather than give Sullivan free reign to undertake the project, the General Assembly ordered,

> That the proprietors of The Enfield Bridge shall again be authorized and empowered to lock the upper part of the Enfield Falls, called Mad Tom and Surf

---

26 "Enfield Bridge Breaks," *Hartford Courant*, February 16, 1900.

27 John L. Sullivan, "Letter on Inland Navigation to the Inhabitants of Hartford and Other Towns on the Connecticut River," *Hartford Courant*, February 14, 1818.

28 *An Act to Incorporate John L. Sullivan and His Associates for the Purpose of Construction Locks, Dams, and Channels at Enfield Falls*, Private Laws of Connecticut, Asser, Vol.1, p. 507.

Bar; provided a majority of proprietors shall, within thirty days from the rising of this present Assembly, pass a vote, and take measures to carry the same into effect without delay; or otherwise, shall, at their option, have a right to subscribe in proportion to their present ownership, for one half the shares in the locks and channels.[29]

Essentially, the General Assembly gave the proprietors of Enfield Bridge Company a choice. Either make the improvements on your own or let Sullivan make the improvements. If they chose the latter, then their franchise rights would be compensated by allowing them to purchase up to one-half of the stock in Sullivan's enterprise. Once again, however, where the average man would perceive that the General Assembly had left the Bridge Company with two options, the proprietors of the Bridge Company determined that they had, in fact, been given three options. The third option was to do nothing, and that is the option the Bridge Company chose. There is no record that any vote to make the improvements was undertaken by the Bridge proprietors, nor is there any evidence that Sullivan ever began to solicit stock subscriptions for his enterprise. The project was again stalled.

The Erie

Although there were several American canals in operation by the time ground was broken for the Erie Canal on July 4, 1817, these were canals of primarily local significance and utility. The Erie was an ambitious undertaking on a previously unknown scale. It would connect the Northeast United States with the Great Lakes, Canada, and the emerging states of the new Mid-West.

New York's Governor Dewitt Clinton expended all of his political capital, and perhaps the capital of many other allies, in wholeheartedly endorsing the Eire Canal project. Clinton, and the canal commission charged with undertaking the project, labored under the never-ending skepticism that the project could be successfully completed and suggestions that "Clinton's Ditch", as his detractors called it, would be a tremendous waste of public funds. Unlike the earlier canals, the Erie was never intended to be financed by private funds. The Erie was conceived

---

[29] Ibid.

as a public works project, the largest any American state had undertaken to that date. The initial legislation authorized an expenditure of up to $5,000,000 to construct the canal.[30]

After securing the state's financial commitment to the project, the first practical problem faced by the canal commission was the fact that there were not enough trained engineers in the country to design the canal. Judge Benjamin Wright, who in addition to his legal training, was an experienced surveyor, was engaged as the Chief Engineer. Wright assembled a small group of ambitious and intelligent, albeit inexperienced, young men to form the corps of engineers who would design the canal. Among this group was Canvass White, who was sent to England in 1817 for one year to observe the English system of canals and return to the United States to share with his comrades as much of the technological underpinnings of the English canal system as his memory and notepads could hold.[31] This "learn as you go" approach to designing the Eire Canal was not ideal, but it was sufficiently effective, in large part due to the character and determination of the men engaged. Over the next fifteen years, Canvass White would be engaged as the Chief Engineer on many American canal projects, including the Union Canal in 1824, the Delaware and Raritan Canal in 1825 and the Lehigh Canal and Enfield Falls Canal in 1827. He was a consulting engineer for the Schuylkill Navigation and for the Delaware and Chesapeake canal.[32]

The second problem faced by the canal commission was the lack of labor needed to build the canal. In the early years of construction, the workforce primarily consisted of German, Dutch, Scottish, and Scandinavian farmers who arrived in the area a generation earlier to establish their farms. But as the pace of construction quickened and malaria, cholera, and other diseases increasingly took their toll on the workforce, the need for foreign workers increased. The political, social, and economic oppression that had been driving young Irishmen from their native country for years coincided with this increasing need, and

---

[30] Peter Bernstein, *Wedding the Waters* (New York, W. W. Norton & Co., 1006), p. 195.

[31] Ibid., p. 192.

[32] *Biographies of Engineers, United States Canals, Supplement to the Annual Report of the State Engineer and Surveyor* (State of New York, 1905), 1170; Canvass White, Raphael, Aaron Craig, Union College, 1996.

many of those immigrants found work on the Erie Canal during the later years of construction.[33]

Even with local farmers and the arrival of additional manpower from abroad, the project could not have been successful without a viable system to manage such a massive workforce. The solution that emerged was the independent contractor system, whereby the canal commission awarded contracts to entrepreneurs who were engaged simultaneously to complete the project in sections. Generally, the excavation work was let out in one-mile sections with the contractor agreeing to complete his section for a uniform price within an agreed-upon period. Often subcontractors were used to complete each section. The independent contractor system had advantages over earlier systems. By pitting contractor against contractor as they bid for work on the canal, costs were effectively contained. The independent contractor system also worked to keep the project on schedule. If one contractor could not perform satisfactorily, he could easily be replaced, and only the section for which he was awarded the contract would suffer.[34]

One of the lessons in canal technology learned by the young engineer Canvass White on his year-long trip to England, was that locks and aqueducts constructed of stone were far superior to those made of wood. Wood was less expensive in the short-term, but its failure rate made it far more costly than stone in the long-term. White learned that hydraulic cement, a mixture that could set and remain firm under water, was required to effectively build stone locks and aqueducts. Unfortunately, no extensive deposit of volcanic pumice, the main ingredient needed to make hydraulic cement, had yet been found in the United States. The engineers made the decision to use stone, but to point the stones with just trace amounts of hydraulic cement imported at great expense from Europe. Unsatisfied with this decision, White took it upon himself to locate a significant deposit of volcanic pumice, also known as trass, within the United States. In this effort he was eventually successful. He went into business with his brother to produce hydraulic cement for use on the Erie and other canals. It is arguable that his contribution of hydraulic cement to the canal build-

---

[33] Shaw, *Canals for the Nation*, p. 38.

[34] Bernstein, p. 206.

ing era was even more valuable than his extensive engineering work.[35]

Spurred on by the unwavering support of its champion, De-Witt Clinton, and driven forward by the perseverance, dedication and creativity of its engineers, the Erie Canal was completed on October 26, 1825. Its immediate success did not go unnoticed by the champions of internal improvement in other localities.[36]

---

[35] Bernstein, p.206; Shaw, p. 38; *Biographies of Engineers, United States Canals, Supplement to the Annual*

[36] *Report of the State Engineer and Surveyor* (State of New York, 1905), 1170; Canvass White, Raphael, Aaron Craig, Union College, 1996.

CHAPTER FOUR

Competition From New Haven

New Haven's Maritime Setbacks

When Reverend John Davenport and a group of London businessmen decided to transplant to New England in 1637, their combined wealth would have made them a welcome addition to any of the existing settlements. Davenport's group, which included Samuel Eaton and Edward Hopkins, was determined to establish an independent colony. These men had made their fortunes as traders. While farming would be a necessity for survival, trading would be the primary source of their anticipated success in the New World.

In the spring of 1638, Davenport's group chose a spot on the North shore of Long Island Sound at the confluence of the Quinnipiac, Mill and West Rivers. This site became the New Haven Colony. Almost immediately, it became apparent that the site was less than ideal for its intended purpose as the hub of a grand shipping enterprise.

Despite its location on the Sound, New Haven's harbor was shallow and posed an impediment to deep draft vessels. Davenport ordered all able-bodied men to wade into the harbor and dig away at the shoals. The results of the effort were unsatisfactory. Not only was the harbor shallow, but of the three rivers that emptied into it, none were navigable for any useful distance, limiting New Haven's opportunities for inland trade.

The New Haven colonists were eager to expand their reach and influence. They quickly purchased from the native Indians a large portion of what is now Southern New Jersey and another parcel on the opposite shore of the Delaware River. Although referred to as the Delaware Plantations, these settlements were not so much intended for agriculture as they were intended to serve as trading outposts for what the colonists expected to be an ever-expanding shipping network. In establishing these outposts, the New Haven colonists ignored the competing claims to the land by Dutch and Swedish interests. This oversight led to such instability and turmoil that the settlements were ultimately abandoned at a total loss to the New Haven entrepreneurs.

With their considerable wealth diminished, the New Haven colonists devised one last "Hail Mary" scheme to establish their stature in the developing world of international trade. In 1646, they pooled their resources to engage a Rhode Island ship builder to produce a cavernous cargo ship. The ship, weighing between 100 and 150 tons, arrived at the end of the year. As she was brought through the Sound to New Haven, all who observed her movement through the water, including the ship's captain, George Lamberton, noted that the ship was "walty." In other words, it had a frightening inability to remain upright, but rather tended to list from side to side at close angles to the water's surface. Undaunted, the New Haven colonists feverishly loaded nearly all of their remaining tradable goods aboard the wobbly vessel for its maiden voyage to London. Several of the colony's most prominent residents planned to make the trip. A successful voyage would replenish their dwindling wealth and, more importantly, demonstrate the colony's shipping capabilities to London investors.

When the departure day arrived, the colonists found their investment hemmed in by an iced over harbor. It would seem that the devout Puritans of the New Haven colony should have perceived the hand of God at work, perhaps to prevent the commencement of an ill-fated journey. But the colonists forged ahead. They cut a three mile channel through the ice and used a tow line to begin their great commercial venture. The "Great Shippe" left New Haven not under the power of her majestic sails, but wobbled off for London, under tow, stern-first.

The ship never made it to London. In fact, the ship was never heard from again until June of 1648. By that time, the remaining colonists were forlorn and had abandoned all hope of seeing the ship or their loved-ones again. In the late afternoon, following a tremendous thunderstorm, word from the harbor quickly spread among the colonists that the ship had been spotted, in full sail, riding a great cloud toward the harbor. The colonists rushed to the harbor. The ship appeared to be close enough that a rock could have been thrown onboard from the shore. Suddenly, the main mast snapped and became tangled with the sails. The remaining masts then snapped and the ship pitched wildly sideways, throwing the occupants into the water. Within minutes, the ship disappeared among the clouds. The apparition was so life-like that the colonists took to their boats to examine the harbor for wreckage but found none. Ultimately, Reverend Davenport determined that the "Phantom Ship," as it became known, was a gift from God to explain the fate of the failed enterprise.[1] Henry Wadsworth Longfellow retold the legend in his poem, "The Phantom Ship."[2]

The Farmington Canal and the Hampshire and Hampden Canal

Even before the opening of the Erie Canal in 1825, the pace of its construction and the prospects for its success kindled a growing enthusiasm for the canal movement. The Erie was not going to be merely a New York success story but a national one. Prominent citizens in cities and states outside New York began to envision a day when a robust system of interconnected internal waterways transported goods and people throughout the developing country at speeds that were previously unimaginable. New canal schemes began to pop up everywhere.

Despite its checkered history in maritime affairs, New Haven was quick to embrace the promise of the canal movement. Proponents rationalized that the lack of a reliable inland transportation route was the cause of New Haven's prior failures. They believed that the prior failures could be reversed by a North to South water route through the heart of New England, with New Haven as the gatekeeper. Such a route was envisioned as an ideal complement to the Erie, especially if a Western link to the Erie could be established. James Hillhouse, the Treasurer of Yale University, was the first to champion the cause. By 1822, a committee

---

[1] David E. Philips, Legendary Connecticut, Tales of the Nutmeg State, The Phantom Ship of New Haven, (Willimantic, CT, Curbstone Press, 1992).

[2] Henry Wordsworth Longfellow, "The Phantom Ship," https://www.hwlongfellow.org/poems_poem.php?pid=87

had been formed to examine the issue.[3] From this committee emerged an ambitious plan.

The Connecticut Legislature adopted the committee's plan in May of 1822 when it passed, "An Act Incorporating the Farmington Canal Company."[4] The canal was to be constructed, "from tide waters of the harbor of New Haven, through this State to the north line thereof, at Southwick, in the State of Massachusetts, passing through the town of Farmington; and also from Farmington up the Farmington River, to the north line of the Town of Colebrook, it being the State line, passing through the town of New Hartford."[5] This Connecticut venture would be coupled with a Massachusetts link from Southwick to Northampton, where the canal would join the Connecticut River. The entire project would be an artery to New England's heartland. Certainly, after two centuries of rivalry between New Haven and Hartford, the proponents were not dismayed that Hartford would be bypassed. The Massachusetts legislature chartered the Hampshire and Hampden Canal Company to provide the northern link in 1823.[6] Massachusetts' involvement in the plan served to widen the rivalry between the prominent citizens of Hartford and New Haven. Because the plan bypassed Springfield as well as Hartford, Springfield tended to side with the Hartford interests. Northampton citizens, who saw their importance increasing as gatekeepers of the northern link, sided with the New Haven interests.[7]

The battle of words and ideas raged on not only in the halls of each State's Legislature, but in the editorial pages of the newspapers of each city. Hartford's prominent traders favored a series of river improvements so that the Connecticut River would be easily navigable to Barnet, Vermont. The river improvement proponents became know as the "Riverites." Those favoring a canal from New Haven to Northampton

---

[3] George M. Guignino, *The Farmington Canal 1822-1847: An attempt at Internal Improvement* (Yale-New Haven Teachers Institute, 1981), UNIT 81.CH.04.

[4] Resolves and Private Laws of the State of Connecticut, An Act Incorporating the Farmington Canal Company, Passed May 1822.

[5] *Resolves and Private Laws of the State of Connecticut, An Act Incorporating the Farmington Canal Company,* Passed May 1822, Section 1, pp. 310 - 318.

[6] *Shaw, anals for a Nation,* p.53.

[7] Drago, *Canal Days,* p.22.

Watercolor depiction of Farmingron Canal vessel. The canal operated its first boat in 1828 and became a highway for trade into central Connecticut. Courtesy, The Farmington Historical Society.

became know in the editorial pages at the "Canallers."

While the optimism of the Canallers seemed justified, there was one significant difference between New York and its support for the Erie Canal and Connecticut and its grant of authority for the Farmington Canal. New York embraced the project as a public enterprise, and was willing to provide financial backing for the project. The Connecticut legislature merely authorized the project without financial support. The Farmington Canal Company was set up as a stock corporation with the right to raise capital by selling $100.00 share subscriptions.[8] Usually, about ten percent of each subscription was paid upon purchase. The balance was paid, when assessed, according to the needs of the corporation.

Erie Canal veteran Benjamin Wright was hired as Chief Engineer. The Farmington Canal was no small undertaking. There is a rise and fall of 520 feet between New Haven and Northampton. The route chosen by Wright encompassed 58 miles in Connecticut and another 20 miles in Massachusetts. Sixty locks were required to overcome the terrain. In addition to the main canal, the anticipated side-cut from Farmington to Colebrook presented more challenges. Despite these engineering obstacles, Wright declared that the cost of construction per mile would be less

---

[8] *Resolves and Private Laws of the State of Connecticut, An Act Incorporating the Farmington Canal Company, Passed May 1822,* Section 2.

than the Erie.[9]

Wright left one important variable out of his calculation. Land acquisition costs were not considered. The entire route crossed private lands. The corporation's charter gave the company eminent domain rights, and a committee was selected to determine the fair market compensation for land taken for canal use.[10] But no amount of compensation could soften the hearts of tough Yankee farmers, who resented their land being taken, especially when it resulted in a division of their prime pasture lands. Several lawsuits were filed by frustrated landowners and these suits tied up needed capital. In addition to lawsuits, the landowner resentment was frequently expressed outwardly in the form of sabotage to the works, which plagued the canal throughout its existence.[11]

Fundraising was not as brisk as expected by the promoters. The Connecticut Legislature, while still unwilling to commit State funds, did lend assistance first by legislating that any dividends on canal stock would be tax free until the company was earning a six percent return on its investment and later by chartering the Mechanics Bank of New Haven upon the condition that it purchase $200,000 in canal stock subscriptions. It was not until 1825 that the company had raised the nearly $500,000 it needed to commence the project. Even then, the company was seriously under capitalized because of the unforeseen litigation expenses and land costs.[12]

Ground was broken in an extravagant ceremony on July 4, 1825. Just as New Haven citizens almost two centuries earlier pressed on as their doomed "Great Shippe" wobbled backward through the ice out of New Haven harbor, they were likewise undaunted when the ceremonial spade used in the groundbreaking ceremony for the Farmington Canal snapped at the handle before turning its first chunk of earth.[13]

To meet its expenses, the company was forced to make assess-

---

[9] Guignino, *The Farmington Canal 1822-1847: An attempt at Internal Improvement*.

[10] *Resolves and Private Laws of the State of Connecticut, An Act Incorporating the Farmington Canal Company, Passed May 1822*, Section 3.

[11] Guignino, *The Farmington Canal 1822-1847: An attempt at Internal Improvement*.

[12] Ibid.

[13] Ibid; Shaw, *Canals for a Nation*, p. 53; Drago, *Canal Days*, p. 22.

ments on its stock subscriptions at a rate faster than anticipated. Many of the subscribers could not keep up with the payments.[14]

Benjamin Wright engaged brothers Davis and Jarvis Hurd. Davis Hurd became chief engineer of the Massachusetts section and Jarvis Hurd rose to chief engineer of the Connecticut section. Out of financial necessity, the Hurds disregarded sound engineering considerations in favor of less expensive alternatives. Wooden locks were substituted for stone structures in many cases. No puddling clay was used to prevent the water from draining through the porous soil. The banks were ill-formed and subject to cave-ins.[15]

Despite the financing difficulties and the engineering concessions, work did progress, albeit at a slow pace. The construction work was performed by contractors who employed independent construction crews. Contracts were awarded for each half-mile section of canal. The use of contractors was the system developed during construction of the Erie. The following advertisement appeared in the *Connecticut Courant* on March 1, 1826:

FARMINGTON CANAL

> Proposals in writing will be received on the first Tuesday of April next, at Bradley's Inn in Hamden, for constructing the Farmington Canal from Cheshire (to where it is already contracted) to New Haven line, near William Love's dwelling house. The line is divided into sections of about half a mile each, and may be further subdivided as shall best suit contractors. Persons making proposals will state separately the price for which they will do the various kinds of work herein enumerated, to be estimated in the following manner, viz.
>
> The gross sum for clearing and grubbing on each section, or part of a section, for which they make proposals:
>
> The price for cubic yard for Excavation;
>
> The price per perch for mason work for Culverts, to be laid in the water line;
>
> The price per perch for mason work of the Abutments of Bridges, dry wall;
>
> The price for wood work of Road Bridges; also for Farm Bridges.
>
> Sealed proposals may be left with me on the line, or at said Bradley's Inn.
>
> James Hillhouse

---

[14] Guignino, *The Farmington Canal 1822-1847: An attempt at Internal Improvement*.

[15] Ibid.; Shaw, *Canals for a Nation*, p. 53.

Superintendent of the Farmington Canal[16]

The contractors completed the work on the Farmington and Hampshire and Hampden Canals with a workforce of predominantly Irish immigrant labor. Working conditions were perilous. The May 21, 1827 edition of *The Hartford Times* contains this item:

> On Monday last, as seven Irishmen were at work on the Hampshire and Hampden Canal, in the vicinity of Westfield, the surface of the earth, which was a considerable height above them, suddenly gave way; and one of their number was buried between four and five feet beneath the falling mass. His comrades were seventeen minutes in digging for their lost fellow, and when discovered, he was apparently lifeless with suffocation, hopes are entertained for his recovery.[17]

In addition to the physical demands and hazards of the labor, the undercapitalization of the project appears to be a further source of worker discontent. James Hillhouse wrote the following letter to the Farmington Canal Company on December 8, 1827.

> Dear Sir,
>
> The Irishmen who worked on the North half of Sec: 62 have completed their job and have a final estimate the balance of which is $428.85 which the men are very anxious to receive – but it is not in my power to pay it having paid out almost the whole of the money which I brought up – I think it very desirable they should be paid if it be practicable to obtain the money – But not being authorized to draw for the money I have requested Mr. Beach who is authorized to receive their estimates to furnish each man with a certificate of the amount due them severally – you know we have thirty days to pay final estimates but it would be desirable that an estimate of this description should not be delayed – you will however do what is practicable and best.
>
> your obt. servant,
> James Hillhouse[18]

The letter refers to the practice of "paying in script" which is pay-

---

[16] *Hartford Courant*, March 1, 1826.

[17] *Hartford Times*, May 21, 1827

[18] Letter of James Hillhouse to Capt. James Goodrich, Dec. 8, 1827. Connecticut Historical Society.

ment in the form of a promissory note for the labor performed. This was a practice despised by the workmen, because it forced them to wait for their money and, in some cases, the money never came at all. Despite the consistent money shortages, by the Summer of 1827, it looked like the entire route from New Haven to Southwick would be opened by the coming fall. Difficulties were then encountered constructing the last lock, connecting the canal with New Haven Harbor. Groundwater poured into the excavation pit faster than it could be drawn out.[19] This problem delayed the opening of the canal until the Spring of 1828, when the canal briefly opened from New Haven to Farmington. The northern terminus was pulled back to Southington after heavy rains caused a major cave-in along the route between Southington and Farmington in September 1828.[20]

At about this time, in an effort to raise morale and restore confidence, the Canal companies invited New York Governor DeWitt Clinton to review the progress of the canal and comment on its potential for success. Clinton's stature was supreme among canal enthusiasts, having championed the cause of the Erie Canal and advanced the project to a successful conclusion. Clinton's enthusiastic endorsement of the project seems to have provided the help the canal needed. The construction setbacks had put extreme financial pressure on the Farmington Canal Company. Not only did they increase construction costs, but they delayed the company's ability to collect toll revenue. Perhaps buoyed by Clinton's visit, the City of New Haven greatly relieved the financial pressure when it borrowed $100,000.00 on its own credit to invest in the canal. The influx of capital enabled the company to settle pending claims, repair any cave-ins, and complete the remaining improvements needed to open the entire length of the canal.[21] The following article from the October 9, 1829, *New Haven Daily Advertiser* was reprinted in the October 13, 1829 *Connecticut Courant*.

The Canal

Arrived, on Wednesday, by the FARMINGTON CANAL, in twenty-two hours from Massachusetts, the fine boat SACHEM, Capt. Owen, with

---

[19] *Springfield Republican*, "The Last Lock on the Farmington Canal," Nov. 14, 1827.

[20] Drago, *Canal Days in America*, p. 23.

[21] Guignino, *The Farmington Canal 1822-1847: An attempt at Internal Improvement*.

passengers from Southwick, Granby, and Simsbury. This is the first arrival through the entire line of the canal, now freely navigable. The boat was greeted with much enthusiasm by our citizens, who displayed, toward the passengers, their usual hospitality with great alacrity and pleasure. This arrival was not needed to restore confidence to this great work. Its stability has been tested by severe storms, and its supply of water by a season of uncommon drought. The canal is now COMPLETED; and, though too late in the season to admit of an extensive carriage of goods this year, it is open for the next, when, at a moderate calculation, the tolls may be expected to exceed fifteen thousand dollars..

Few expensive works have been accomplished, through so many difficulties and discouragements as this Canal Company has encountered and overcome. The obstacle interposed by the active enemies of the canal, the disheartening influence of certain cold friends, "damning with faint praise," unavoidable disasters, all have not depressed the spirit of enterprise which projected, and has finally completed, a work, the benefits of which are not yet half appreciated.

It is not now the time to look back for those causes of distrust, - they might all be traced to the invidious jealousy of a few who take no joy in the general prosperity of the city, - which have impeded the work, and which at one time almost destroyed the credit of the company. The doubts are now dissipated, the hopes are confirmed, and credit is restored, and it is pleasanter to look forward in anticipation of future good, than back upon past evil.

It has been said, and, we have no doubt, with entire truth, that one house has transported 500 tons of cheese to this market already. Boats with full freights are constantly arriving and departing. The activity, considering the unfinished state of the canal, has been great, through the summer – and the receipt of tolls has exceeded the most sanguine expectations, having been, we believe, sufficient to defray the ordinary expenses and repairs. The number of boats another season, will be greatly augmented, many are already building and many more are ordered. The prospect is certainly encouraging.[22]

Given the difficulties experienced by the company prior to the opening of the canal, it is no surprise that the tone of the article is somewhat muted. Confidence in the enterprise is clearly expressed yet one can sense some lingering, unexpressed worry on the part of the author. The author leaves nothing unexpressed when it comes to his enmity toward the "active enemies of the canal." Cleary, the animosity was real. Fortunes, either those to be made or those to be lost, were on the line.

---

[22] *New Haven Daily Advertiser* reprinted October 13, 1829 in *The Connecticut Courant*.

CHAPTER FIVE

The River Improvement Scheme

The Farmington Canal and the Hampshire and Hampden Canal projects posed a calculated, serious threat to the prosperous traders of Hartford. An examination of a map of the present city of Hartford reveals many streets named in memory of the Hartford businessmen who were early proponents of river improvement. William Imlay, David Watkinson, H. L. Ellsworth, Eliphalet Averill, Charles Sigourney, Joseph Pratt, John Russ, James Goodwin, and Daniel Buck were among the Hartford "Riverites." Many of these men accumulated their fortunes in trading "West Indian" goods. Hartford was a flourishing market whose citizens eagerly awaited shipments of rum, molasses, spices, and sugar from exotic Caribbean ports. In addition to West Indian goods, international trade in such products as wine from Madiera, gin from Holland, and lace from Ireland was on the rise, and Hartford merchants all stood to profit. The local newspapers announced the arrival of each ship, described its cargo, and identified the merchant retailing the goods.

Many of these traders met at the City Hotel during the first week in January 1824 to devise a response to the New Haven threat.[1] From these meetings evolved a comprehensive scheme of river improvement. The men were merchants, not sailors, but their knowledge of the shipping industry familiarized them with emerging trends in transportation. To retain their historical advantage over New Haven, their river improvement scheme needed to offer a competitive advantage over canal transportation. They needed to look no further than the Hartford riverfront for the answer. The competitive

---

[1] Henry Erving, *The Connecticut River Banking Company 1825-1925* (Hartford, The Connecticut River Banking Company,1925), p. 51.

advantage they sought was found billowing in the sky above the city docks.

Steam Power

John Fitch, a native of South Windsor, Connecticut is credited with being the first man to propel a vessel through water using steam power. His odd-looking craft, with a series of upright oars on each side tethered to reciprocating steam driven pistons, was successfully operated by Fitch on the Delaware River in 1785. Samuel Morey of Oxford, New Hampshire, further developed the technology by incorporating the side paddle wheel into the design in 1792. Many of his experiments were conducted upon the Connecticut River.[2]

Robert Fulton, often erroneously credited with inventing the steam ship, can justifiably be credited with popularizing the new technology. His steamer, *The Fulton*, drew thousands of Hartford citizens to the waterfront when it made its first trip up the Connecticut to Hartford on May 11, 1815. The steamboat *Enterprise* was the first to offer regular trips between Hartford and Old Saybrook in 1819. The Connecticut Steamboat Company was chartered by the Connecticut General Assembly later that year. Interstate steamboat transportation was hindered because the New York General Assembly had granted Fulton a monopoly over the use of steamboats in New York waters. But with a challenge to that monopoly pending in the United States Supreme Court, the Connecticut Steamboat Company commissioned the construction of the *Oliver Ellsworth* in New York in 1823. The ship was mammoth and elegant - 112 feet long, 24 feet wide, a gentleman's cabin and a ladies' cabin each with 16 berths, a dining room, promenade deck and a 44 horsepower engine. When Justice Marshall, writing for the majority, overturned Fulton's monopoly on March 2, 1824, the way was opened for regular steamship service between Hartford and New York, and the *Oliver Ellsworth* was immediately put to that service.[3]

The transportation revolution brought about by the harnessing of steam power and its application to deep water navigation was not missed by the Hartford merchants who met in early 1824 to devise an upriver transportation scheme to keep ahead of their rivals in New Haven. The current state of steam technology did not provide the full answer, howev-

---

[2] Ibid, p. 75.

[3] Ibid., p. 85.

er, as upriver transportation required vessels that could operate in shallow water, and vessels capable of passing through the narrow locks of upriver canals. The feasibility of such vessels had been advocated as early as 1818 by John L. Sullivan, superintendent of the Middlesex Canal, who faced similar challenges in transporting products from the extension of the Middlesex canal up the Merrimack River in New Hampshire. In fact, Sullivan intended to introduce the developing technology to the Connecticut River when he gained legislative approval to succeed the Enfield Bridge Company in the effort to bypass the upper Enfield Falls in 1818.[4] Sullivan never pursued this project because as part of the approval the legislature gave the owners of Enfield Bridge Company financial control over Sullivan's enterprise in an effort to protect their franchise rights. The short-sighted Reynolds and his successors were never interested in anything more than the bridge franchise that came with their canal building rights.

The plans devised by the Hartford merchants were ambitious. A fleet of steam propelled tow-boats would be developed to tow barges and scows laden with goods up and down the river. The Enfield Falls, both upper and lower, would finally be bypassed by a lateral canal. Upriver canals would be purchased and improved where necessary to provide an uninterrupted transportation chain. The entire river route from Hartford to Barnet, Vermont and then northeast to Lake Memphremagog on the Canadian border, would be surveyed to determine the feasibility of the plan and to identify any further needed improvements.

While the Hartford merchants had a clear vision and were single-minded in furtherance of the plan, they lacked leadership. They needed a man who was not just savvy in mercantile affairs, but a facilitator who knew his way around the halls of the State Capitol, who understood corporate dynamics and corporate finance and who had enough enthusiasm to see the plan through. They enlisted the aid of Alfred Smith, a prominent Hartford lawyer. Smith was a Judge of the County Court. He served as the representative of the City of Hartford in the State legislature and was familiar with the legislative process. As a native of South Hadley, Massachusetts, home of the South Hadley Canal, he also maintained strong connections to upriver navigation interests. Alfred Smith took the leadership role in forming The Association for Improving the Navigation of the Connecticut River. The Hartford merchants were appointed as Directors of the Associ-

---

[4] *An Act to Incorporate John L. Sullivan and His Associates for the Purpose of Construction Locks, Dams, and Channels at Enfield Falls, Private Laws of Connecticut*, Vol.1, p.507.

ation and Alfred Smith was appointed as its President.[5]

The First Survey

Following formation of the Association, a committee was formed, "To examine and survey the obstacles in the Connecticut River above Hartford, also to enquire into the most practicable method of improving said navigation, with general powers to act, in behalf of said Company, etc."[6]

A member of the committee traveled to Albany to secure the services of an experienced and competent engineer to conduct the survey and make recommendations that would address the committee's purposes. The effort met with some delay resulting from the relatively small number of qualified engineers in the field, and the ongoing duties with regard to the completion of the Erie Canal of those who were capable of the work. Canvass White of Troy, New York was eventually engaged to perform the survey. White assembled a group of assistant engineers and surveyors. The group initially undertook a general examination of the Connecticut River and the existing canals and improvements between the Enfield Falls and Miller's Falls. The group then began a detailed survey of the Enfield Falls. The detailed survey was a prerequisite to making recommendations and estimates for potential methods of circumventing the rapids. White was recalled to New York during this period for over one month to attend to an illness within his family, but the team he assembled was sufficiently skilled to proceed in White's absence. When White returned, surveying activities above Connecticut were suspended so that the team could gather to review their findings to date and begin to formulate their recommendations. The group had spent sufficient time on the upper river to calculate the depth of each of the falls between Well's River and Hartford and determined that the total fall between those two points was 371 feet.[7]

On October 23, 1824, White presented the survey, recommendations, and estimates to the committee. He opined that the cheapest method for improving the navigation of the rapids would be to erect a

---

[5] *Two Reports Made by Committees Appointed by the Directors of the Association for Improving the Navigation of the Connecticut River Above Hartford, January 1825* (Hartford, 1825).

[6] Ibid.

[7] Ibid.

THE RIVER IMPROVEMNENT SCHEME 65

dam, across the entire river just below the lower rapids, with an embankment, short canal and locks just below the dam. He quickly excluded this option as it would have inundated the floodplain areas above the dam. The land that is now Main Street in Windsor Locks and Bridge and Main Streets in Warehouse Point would have been under water. Further, a complete damming of the river was not permitted by the corporate charter. Instead, White suggested three possible configurations for lateral canals to bypass the falls.

The first configuration is closest to the canal that would eventually be constructed. He proposed a substantial wing dam on the west side of the river at the head of the upper falls, which would be continued parallel to the west bank of the river to a suitable point for a guard lock. An embankment of earth would then run from the east wall of the guard lock parallel to the west bank of the river to the northern tip of King's Island. From the southern tip of Kings Island, the embankment would head back toward the west bank, and then continue south, parallel to the west bank, to the end of the lower falls. A short canal could then be excavated on land along the bank of the river to a convenient point for locking down to the river. The man-made embankment, and King's Island itself, would become the eastern wall of a canal. White estimated the cost of such a plan to be $97,058.00.

The second configuration was to construct a false embankment parallel to the west bank of the river to the foot of the upper falls, and then to lock down into the slack water between the falls. A second channel would then commence near the head of King's Island with locks for re-entry to the river after the end of the lower falls. A guard lock would need to be included with each section as well as two lock systems to return boats from the shore channel to the river. White estimated the cost of this configuration to be $84,752.00.

The third proposed configuration involved a canal on the east side of the river commencing above the upper falls and continuing south until it passed the lower falls. The canal would incorporate a combination of shore channel sections and excavated canal sections as the terrain permitted. The cost of this configuration was estimated to be $87,203.00.[8]

Since White's survey was cut short by winter weather, he did not have the time to take the necessary measurements that would have en-

---
[8] *Report of the President and Directors of the Connecticut River Company, January 1826.*

abled him to suggest detailed improvements or cost estimates for upriver navigation. He did identify the upriver obstacles and render the opinion that with some moderate improvements and by utilizing steam towboats, a controlling depth of three feet was both sufficient and attainable.

The committee communicated its findings, including the information provided by Canvass White, to the directors of the Association on November 18, 1824. Following this meeting, the Directors passed three resolutions: 1.) to organize the corporation pursuant to the charter received from the state legislature the previous May; 2.) to seek Federal assistance to secure a detailed survey of the entire Connecticut River Valley; and 3.) to organize a meeting with representatives from each of the towns in or near to the Connecticut River Valley which might benefit from the contemplated river improvements.[9]

Incorporation of the Connecticut River Company

The ambitious resolutions adopted by the Association indicated that the river improvement plan had gained enough momentum and support that the Riverites felt comfortable taking the plan to the State Legislature. Opposition from New Haven and Farmington valley interests was fierce, but in May of 1824, the Connecticut General Assembly approved a charter to organize the Connecticut River Company and the Directors of the Association For Improving the Navigation of the Connecticut River resolved to organize the corporation pursuant to that charter on November 18, 1824. The resolution was approved by the full membership of the Association shortly thereafter. At a general meeting of the corporation held on January 2, 1826, Ward Woodbridge, David Porter, William Ely, Daniel Buck, Henry Kilbourn, Thomas Brace, and George Beach, were elected the Directors of the corporation. Alfred Smith was appointed President, James Goodwin Secretary, and Daniel Buck Treasurer.[10]

The legislature empowered the corporation to raise up to $500,000.00 in capital by the sale of common stock subscriptions in the amount of $100.00 per share. The payment for each share of stock was to be made in installments. The amount of each installment and the frequency of the installments would be determined by the directors. If

---

[9] Ibid.

[10] *Report of the President and Directors of the Connecticut River Company, January 3rd, 1826.*

a shareholder failed to keep up with his installments, his shares would be forfeited and sold at public auction. After deducting the delinquent installments and the costs of the sale, any proceeds of the sale would be returned to the defaulting shareholder.

The corporation was authorized to widen the channel of the river from Hartford to the Massachusetts line by removing material and obstructions in the river and by building any "wharves, piers or hedges" in the river or upon the banks as necessary, to lock the falls at Enfield, to make channels in the river to aid the locks, to construct a canal on either bank of the river, to construct any needed dam or dams to aid entering or leaving the locks provided they did not prevent the convenient passage of boats or logs in the river or obstruct the passage of fish, and to procure and possess steam boats necessary to increase commerce on the river.

Alfred Smith, President, Connecticut River Company.

The legislature further granted the corporation the power of eminent domain. The corporation could take possession of adjacent lands, water or streams needed to aid in the construction of the canal. If such property was privately owned, and fair compensation could not be agreed upon between the parties, a committee was established to determine the amount. If the property owner was still not satisfied, he was provided with a right to appeal to a committee of three "disinterested freeholders" appointed by the County Court.

As to the upriver improvements north of the Connecticut border, the Connecticut legislature had little authority, but did authorize the company to purchase and hold the stock of any incorporated upriver canal or lock company.

The corporation was given five years to complete the improvements. Upon completion, the corporation was empowered to collect a toll of $0.50 per ton of cargo for every boat going up the river, whether utilizing the canal or ascending the falls. For empty boats or boats with less than two tons of cargo going up the river the toll was a flat $1.00. Boats and rafts descending the river would only have to pay a toll if they actually utilized the canal. If so, the toll was $0.17 per ton of cargo. The tolls were to remain in effect until the company's earnings were more than eight percent above their annual operating expenses, at which point the tolls would be reduced.

Three independent commissioners were appointed to provide oversight of the enterprise. They were to perform the role of a modern-day public utility commission. They could regulate toll rates and they could suspend the right to collect tolls in the event that the improvements were not properly attended or maintained.

The most unique and forward-looking aspect of the corporate charter was the provision allowing the Connecticut River Company to acquire any adjacent mills or manufacturing establishments or any land adjacent to the locks, canal and dams where a mill could be established. This provision indicates that, from the outset, the incorporators recognized that the canal design might present opportunities for the production of water power, and the opportunity to supplement toll revenue by distributing water power to local mills. This insight on behalf of the incorporators would prove to be extremely beneficial in the ensuing years.[11]

The Vermont Convention

The river improvement scheme was a regional plan. The authority of the Connecticut legislature only extended to the improvements that would take place in Connecticut. Farmers, manufacturers, and traders throughout the entire Connecticut River Valley would have to be persuaded as to the feasibility and benefits of the plan. The State legislatures of Massachusetts, Vermont, and New Hampshire would have to authorize the portion of the project that involved their jurisdictions.

To that end, Alfred Smith undertook an aggressive and thorough public relations tour throughout the valley. The editorial pages of the valley newspapers quickly embraced the movement. *The Greenfield Gazette* proclaimed:

---

11 *Resolves and Private Laws of the State of Connecticut*, p. 72.

...should it be successfully completed, (and it is believed by many, that it may and will be) the vast benefit that would accrue from such an improvement, by the increase in population and wealth, can hardly be imagined.[12]

The Bellows Falls Examiner declared that the completion of the improvements, "would be a proud day for the western part of New Hampshire and the eastern part of Vermont."[13]

Smith's efforts culminated in a heavily attended convention in Windsor, Vermont. Almost every river town held town meetings in anticipation of the convention to appoint delegates to represent the interests of its citizens. The convention, held in late February 1825, registered 215 delegates and was attended by an additional 200 interested parties. *The Connecticut Courant* reported on March 1, 1825, as follows:

> The Convention was composed of gentlemen of the first order, in point of talent, wealth and respectability, of the towns they represented. The discussions were conducted with perfect order and decorum, and the display of deep research, extensive intelligence, sound reasoning, and eloquence withal, would have done credit to any legislative body.[14]

Moses P. Payson of Bath, New Hampshire, was appointed President of the Convention. Payson served as a member of the New Hampshire State Senate from 1804 to 1816. Norman Williams of Woodstock, Vermont was appointed Secretary. Williams served in the Vermont State Senate 1854 and 1855. He was a member of the committee which oversaw the construction of the new State House in Montpelier. Carlos Coolidge from Windsor, Vermont was appointed Assistant Secretary. Coolidge was born in Windsor, Vermont, in 1792 and died there in 1866. He practiced law in Windsor for fifty-two years. He was a representative in the State legislature from 1834 until 1837, and again from 1839 until 1842. He was elected governor of Vermont from 1848 to 1850, and senator from 1855 till 1857. Clearly, the convention captured the interest of the most distinguished and influential men in the Connecticut River Valley.

There is some indication that the convention was much less a free interchange of ideas as it was a pep rally among men who had already

---

12 As reprinted in *The North Star* (Danville, Vermont), Tue, Feb 15, 1825.

13 Ibid.

14 *The Connecticut Courant*, March 1, 1825.

been convinced as to the value of the plan. It is likely that is all that Smith intended. The convention was perceived as a great success and an overwhelming endorsement for the cause of river improvement. Strong advocates were now in place and available to support the plan in legislatures of each state.

The Second Survey

Perhaps the most significant resolution to come out of the Vermont Convention was a petition to the United States Congress, emphasizing the national interest in improved navigation of the Connecticut River. Little more than a decade earlier, the British navy's total disruption of coastal navigation during the war of 1812 left the United States military desirous of establishing more reliable inland water routes. The petition to Congress sought federal assistance in conducting a survey of the entire inland route. Alfred Smith immediately went to Washington seeking support for the request and was successful in convincing the War Department to conduct a survey of the Connecticut River from Long Island Sound to Lake Memphremagog. The Connecticut Legislature granted the Connecticut River Company's request that Smith be appointed a commissioner to liaison with the War Department surveyors. The last thing the Company wanted was a survey that ran at odds with the Company's plans. Smith's oversight was thorough. He personally accompanied the government surveyors on their route.

The survey team first concentrated its efforts on identifying two potential routes for canals that would connect the Connecticut River at Barnet, Vermont to Lake Memphremagog, which straddled the border of the United States and Canada. There is no available record of who led this survey on behalf of the War Department and no published copy of the survey was ever produced. The pace of the survey frustrated Alfred Smith and the directors of the Connecticut River Company. When it became clear that the War Department survey would not be completed that year, and that no progress at all had been made below Barnet, Vermont, the ever-eager Connecticut River Company independently engaged renowned Erie Canal Engineer Holmes Hutchinson to survey the route from Barnet, Vermont to Hartford. Hutchinson hired Canvass White as his Assistant Engineer for this project. In June of 1825, Hutchinson, White, two surveyors, and several others traveled to Barnet to begin the survey.

Hutchinson completed his work in December of 1825, and immediately presented his survey to the Connecticut River Company directors. He determined that the entire distance, by water from Barnet, Vermont to Hartford was 219 miles with over 420 feet of descent. He proposed that by incorporating the existing locks and canals and constructing some additional improvements, a controlling depth of three feet could be maintained. The finished project would involve seventeen miles of canal and forty-one locks. Considering that the already existing improvements consisted of eight miles of canal and twenty-eight locks, the additional improvements were considered modest and attainable, even taking into account the recommendation that some of the existing locks be rebuilt and enlarged to assure a uniform size. The Riverites were encouraged by Hutchinson's survey and report.[15]

In addition to the measurements and estimates they provided, the endorsement of the river improvement project by such esteemed engineers as Holmes Hutchinson and Canvass White provided instant credibility to the plan. These men were pioneers of American civil engineering and well-known voices to the promoters of the emerging American canal network.

Holmes Hutchinson was born in Port Dickenson, New York in 1794. He became an engineer on the Erie Canal, under Chief Engineer Benjamin Wright in 1819. He remained in the position until 1835, when he was named chief engineer on the Erie. He oversaw the first enlargement of the Erie, and was personally responsible for the design of the enlarged locks during his service. He surveyed the routes of many American canals, including the Erie, Champlain, Oswego and Chemung. He was involved with the construction of the Cumberland and Oxford Canals in Maine and the Blackstone Canal in Rhode Island and Massachusetts. With the emergence of railroads, his efforts turned to that technology, and he became a director of the Utica and Syracuse Railroad and the Syracuse and Oswego Railroad before his death in 1865.[16]

Canvass White was born in Whitesville, New York, near Troy in 1790. He was frequently ill as a child and his parents sent him to Russia on a merchant vessel hoping the salt air would improve his health. He returned in 1812 to find the country at war. He formed a volunteer militia

---

[15] *Report of the President and Directors of the Connecticut River Company.* Originally Published by Philemon Canfield, Hartford, 1826.

[16] New York State, *Biographies of Engineers, United States Canals, Supplement to the Annual Report of the State Engineer and Surveyor, State of New York,* 1905, p. 1157.

company, was commissioned a Lieutenant, and severely wounded while participating in the capture of Fort Erie. In 1816 he was engaged by Benjamin Wright to assist in the survey of the final route for the Erie Canal. At DeWitt Clinton's request, he traveled to England, were he spent nearly a year examining more than 2000 mile of English canals, locks, and aqueducts. He made detailed drawings of the English canal system. He acquired many of the latest surveying tools during his trip.[17] Upon returning to America, he applied what he learned regarding the composition of English hydraulic cement to seeking similar materials in the United States. He obtained a patent for his American-style hydraulic cement, and with his brother formed a company to produce the material. This innovation solved the need of the growing American canal movement for a material that could quickly set underwater as it bound together the stones of lock chambers and aqueducts. White was also in heavy demand as a consultant, surveyor, engineer and supervisor on many of America's canals, including the Schuylkill and the Lehigh in Pennsylvania and the Delaware and Raritan in New Jersey. His consistently poor health led to an early death at the age of forty-four in 1834.[18]

Mill Sites and the Provision of Water-Power

In furtherance of the far-sighted corporate charter provision which permitted the Connecticut River Company to acquire mill sites along the canal route, on April 29, 1825, the board of directors adopted the following resolution:

> That David Porter and William Ely be a committee to examine the different sites on both sides of the Connecticut River suitable for mills in case of a canal being made by Enfield Falls.

Their favorable report demonstrates that the directors were aware of the possibility of utilizing the improvements not merely as an aid to transportation but for the furnishing of power.[19]

---

[17] Bernstein, *Wedding of the Waters*, p. 192.

[18] New York State, *Biographies of Engineers, United States Canals, Supplement to the Annual Report of the State Engineer and Surveyor, State of New York*, 1905, p. 1170. See, Canvass White, Raphael, Aaron Craig, Union College, 1996.

[19] H. Holmes Hutchinson, *Report of the President and Directors of the Connecticut River Company, January*

The *Barnet*

With rising enthusiasm, and with the prospects for a river improvement scheme that would open the heart of New England to steam-borne transportation looking more and more promising, the Connecticut River Company devised a promotional plan that, if successful, would bring support for river improvements to a fever pitch. It is one thing to propose an ambitious scheme, it is quite another thing to actually demonstrate it. During the summer of 1825, the directors of the Connecticut River Company resolved, "That a steamboat be forthwith procured, or built, adapted to the navigation of the Connecticut River above Hartford and that the directors carry the same into effect."[20]

Sketch of the steamboat Barnet.

Possibly due to financial limitations, construction on the boat did not begin until August 22. 1826. On October 5, 1826, the directors resolved, "That the steamboat now building in New York for the Connecticut River Company, be called the *Barnet*."[21] The name, of course, was chosen to reflect the hoped-for northern terminus of the vessel's range. The *Barnet* was launched in New York on October 2, 1826, and her boilers and engines put through a series of tests and refinements. Not

---

1826., Records of the Office of Governor, 1820-1858, State Archives Record Group No. 005.

[20] Erving, *The Connecticut River Banking Company*, p. 76.

[21] Ibid., p. 101.

willing to risk testing the ship's power upon the unpredictable waters of Long Island sound, the *Barnet* was towed to Hartford by the *MacDonough*, one of the largest vessels in Hartford's growing steamship fleet. The ship arrived in Hartford on the 15th of November. Residents of the upriver towns eagerly anticipated the ship's arrival. The November 8, 1826 *Springfield Republican* notes, "We understand that the Steam Boat Barnet may be expected to arrive at this place by the middle of next week."[22]

Due to the lateness of the season, consideration was given to putting off the *Barnet's* maiden voyage until the spring. Should she encounter ice along the route, the prospects for a successful trip would be greatly weakened. Anticipation of the trip was so high, both by the Riverites hoping for success, and the Canallers, not so quietly desiring failure, that the decision was made to proceed upriver at once.[23] On November 17, 1825, the *Barnet* steamed north from Hartford under her own power. Windsor Locks industrialist and historian Jabez Haskell Hayden, then a fifteen-year-old boy growing up in the Hayden Station Section of Windsor, recalls the *Barnet's* maiden voyage in his *Historical Sketches*, published in 1900:

> When the news reached Hayden's that the steamboat, *Barnet*, was on her way up from Hartford, a lad of my own age and myself took our guns and powder horns and hastened to the river to salute the first steamboat to come above Hartford. The boat was in sight a mile below, being saluted by musketry from both sides of the river, and the sound of the exhaust steam from her low pressure engine equaled the report of the musketry. When the boat reached us we opened fire, and the crowd of men and boys cheered themselves hoarse, and we loaded and fired until we had exhausted our stock of powder. One man walked some distance along the shore and said the boat went just as fast as he could walk.[24]

The *Barnet* reached Warehouse Point as nightfall came. That evening the taverns were lively with excitement. Several falls-men were recruited to help pole the *Barnet* over the falls the next morning.

The morning of the eighteenth greeted the *Barnet* with a stiff

---

[22] *Springfield Republican*, November 8, 1826.

[23] Eving, *The Connecticut River Banking Company*, 107

[24] Hayden, *Historical Sketches*, p. 18.

THE RIVER IMPROVEMNENT SCHEME

breeze from the northwest, hardly ideal for a north-bound vessel. The crew was joined by the falls-men with their setting poles and the *Barnet* set-off against the lower rapids. The boat put up a mighty effort but could only make it to the point of the current railroad bridge before losing ground. Not willing to overtax the boat's new engine, the crew determined to return to Hartford to await better conditions.[25]

News of the unsuccessful attempt to overcome the falls quickly got back to the Canallers in New Haven and Northampton and the editorial pages in the newspapers of those towns wasted no time before reveling in the apparent failure. The Hartford and Springfield newspapers were just as eager to defend the effort. *The Springfield Republican*, of November 22, 1826, contains this account of the first effort to ascend the falls:

> The experiment, of towing the *Barnet* over Enfield Falls, for the purpose of navigating the slack water pools above, was tried on Saturday last, but failed in consequence of a stiff northwester which prevailed throughout the day. Should the present pitch of the water continue, and a good south wind spring up, we understand, the attempt will be renewed, and in the estimation of those particularly conversant with the river, with every prospect for success. Owing to the lateness of the season, however, we hardly dare to hope to see so welcome a visitant.[26]

The same edition contains this bristling admonition:

> *The New Haven Register* says that the Steam Boat *Barnet* 'made several unsuccessful attempts to ascend Enfield Falls.' The boat made but one unsuccessful attempt, and the reason for the failure was given. Stick to the facts, gentlemen.[27]

The Northampton Gazette regaled in the misadventure as follows:

> Never was a stranger of distinction more anxiously and delightfully anticipated. The good LaFayette would have been gratified with the honors paid the *Barnet*. The newspapers fortold and announced her. The people on the river were in an uproar. At Springfield the hubbub was complete. The Streets were thronged, sentinels were placed upon the bells, the can-

---

[25] Eving, *The Connecticut River Banking Company*, p.108.

[26] *The Springfield Republican*, November 22, 1826.

[27] *The Springfield Republican*, November 22, 1826.

non were in readiness for a salute, and a public dinner was ordered for the expected guests and numerous citizens. Besides passengers from Hartford, gentlemen from Springfield and other places on the river went down to 'take passage' – and one and all made ready their throats for the huzza of gratulation and triumph. Alas, for the vanity of human wishes! The Barnet maintained the reputation she had acquired. At the very first obstacle at Enfield Falls she stuck fast. Exertions were made to haul her up. Captains, passengers, and crew lent their aid – and two or three flat bottomed boats, manned with poles, labored to tow the Tow Boat along – but all in vain. The passengers finished their 'pleasant excursion' at Warehouse Point, and got home as they could, and the Barnet was put back to Hartford, towing 'four boats, three of them loaded,' down the river. There we understand she is to winter. Thus commenced, 'the first of a series of improvements.' Here ended 'the first trip up the Connecticut River.'[28]

In fact, the first trip up the river had not ended. The November 29 *Springfield Republican* reported:

The Steam-Boat arrived!

We have only time to mention the fact that the Steam Boat *Barnet* arrived at Springfield yesterday afternoon, about four o'clock; and was greeted by our citizens with every demonstration of gladness. The President of the Connecticut River Company, and two or three other gentlemen were passengers. We understand it is the intention to navigate the boat farther up the river.[29]

The Springfield Republican picked up the story the next week in its December 6, 1825 edition:

We have heard of the passage of the steam boat Barnet as far up the river as Greenfield. It passed the locks at South Hadley in fine style, and was welcomed by the inhabitants with tokens of joy. At Northampton and Greenfield also, we understand she met with a welcome reception. The season has so far advanced that it is not expected she will proceed farther up than Bellow Falls. The passage of this boat up the River, has not only silenced the opponents of the project, but has seemed to satisfy all, even the most doubting, of the practicability of navigating this River by steam. If so much can be accomplished in the present state of the river, what may we not expect when the contemplated improvements take place.

---

[28] *Northampton Gazette*, reprinted in the *Springfield Weekly Republican*, December 06, 1826.

[29] *The Springfield Republican,* November 29, 1826.

P. S. We have since learnt, that the boat was in the third lock of the canal at Montague, at 5 o'clock on Monday, on her way up to Brattleborough.[30]

The Connecticut Courant reported the story in its December 4, 1826 edition as follows:

> The Steam-Boat *Barnet*, recently built in New York for the navigation of Connecticut River above this place, which arrived here on the 15th, started on Monday of last week on a trip up the river. She had previously made an unsuccessful attempt to pass the Enfield Falls and returned to Hartford, and this circumstance had called forth a vast deal of wit and good feeling from some of our neighbors. To correct the numerous misstatements that are abound respecting this boat we avail ourselves of the following concise statement of facts from the *Mirror*.
>
> The *Barnet* – Some response seems necessary to the many statements collected and commented upon regarding this boat, and the following are the facts. It is true that she was towed by the MacDonough a good part of the way from New York to Hartford. It is also true that her first attempts to pass the Enfield Falls were unsuccessful. These two facts came to be highly relished by certain folks from whom better feelings, if not more correct statements, were expected. As to the towing of a small boat of sixty odd feet keel, not intended for navigation of the Sound, with a new engine that had never been tried before, it is useless to talk. Enfield Falls (and the gentlemen who rejoice over her failing to pass them know it well enough) were never intended to be navigated until the contemplated improvements should be made. The wind and the tide, the misfortune of meeting a heavily laden boat, and other obstacles prevented her from going up the falls the first time, and she returned to Hartford. She has now passed the falls without difficulty, and reached Springfield in very good season. She arrived there at 4 o'clock on Tuesday last, and, thanks to the hospitality of the good folks there, was received with true neighborly kindness…The boat answers every expectation of those who were sufficiently acquainted with the difficulties to make sure and rational calculations, -- or, in other words, of those who have been industriously employed on the subject of river navigation.[31]

When the *Barnet* rounded the bend as she came into Bellows Falls, Vermont she was met with a fifty-gun salute fired from a cannon on the village green. Church bells pealed. Alfred Smith and the two captains of

---

30 *The Springfield Republican*, December 6, 1826

31 *The Connecticut Courant*, December 4, 1826.

the *Barnet*, Captain Strong and Captain Palmer, were met by throngs of river valley citizens. A large banquet was held that evening, and Smith and the two captains were joined by the officers from the Windsor, Vermont convention, and leading citizens of Bellows Falls and surrounding river towns. Following the dinner, the *Barnet* and her passengers were toasted at length, with each toast followed by a boom from the cannon on the green. Said one, "To the enterprising, energetic, and persevering inhabitants of Hartford, -- We have long been their friends, we shall soon be their neighbors!" Smith responded to the cheers as follows; "To the citizens of New Hampshire and Vermont, -- Enterprising, liberal and intelligent, may they be prospered accordingly!"[32]

Rare mention was made of the *Barnet* after completing its historic maiden voyage. One report is that the *Barnet* was present towing the "Safety Barge Lady Palmer with a large party to witness the opening of the canal at Enfield Falls," in November 1829.[33] If that report is accurate, then it could not have taken place without first having undergone substantial repairs. The records of the Connecticut River Company contain these two entries on November 3, 1827:

> Voted, That the funeral charges of Mr. Joseph Groumly, who was unfortunately scalded to death by the bursting of the boiler in the steamboat *Barnet*, be defrayed by the Connecticut River Company.
> and
> That M. W. Chapin, or Joseph Pratt, be authorized to pay the charges referred to in the preceding vote, and also to pay off the sum employed to bring the said Steamboat *Barnet* from New York to Hartford.[34]

The Vermont Statesman reported on November 14, 1827:

> The boiler of the Steamboat *Barnet* burst on Friday, the 15th inst. When she was off Milford in Long Island Sound, on her passage from New York to Hartford. Mr. Grumby, a pilot, of Saybrook, was killed. There were no passengers in the boat. The *Barnet* was built for the purpose of navigating the Connecticut River, and has a wheel in her stern.[35]

---

[32] Erving, *The Connecticut River Banking Company*, 117.

[33] *Connecticut River, Quinnehtuk, The Indians Called It, Proceedings of the History of the Pocumtuck Valley Memorial Association*, 1901.

[34] Erving, *The Connecticut River Banking Company*, p. 131.

[35] *Vermont Statesman*, Castleton Vermont, November 14, 1827.

Several newspapers printed similar accounts.[36] *The Vermont Patriot* attributed to the *Hartford Times* a report that the damage done to the *Barnet* by virtue of the explosion was minor and that the explosion was caused by "the parting of three or four of the rivets."[37]

Despite the early difficulties and later tragedy, the *Barnet* served her purpose well. The boat demonstrated the practicability of upriver navigation by steam. Her progress had been followed with great interest and her success celebrated in every river town she reached. Hayden reveals that she surmounted the falls by lashing a flat-boat to each of her sides and utilizing thirty falls-men to propel her up the Enfield Falls.[38] But as the editors of the Riverite-leaning newspapers pointed out, she was never intended to overcome the falls. The promotional scheme was a rousing success. Aside from the righteous snickering of the Canallers, everybody with an interest in the cause of river improvement was calling for the project to move forward. A letter to the editor of *The Hartford Times* published on December 19, 1826 asked the next logical question, "Has not the time now arrived for locking the Enfield Falls?"[39]

---

[36] T*he Litchfield County Post*, November 15, 1827; *Vermont Republican and Journal*. Windham, *Windsor and Orange County Advertiser*, November 10, 1827; *Burlington Weekly Free Press*, November 16, 1827.

[37] *Vermont Patriot,* November 26, 1827.

[38] Hayden, *Historical Sketches,* 29.

[39] *The Hartford Times,* December 19, 1826.

CHAPTER SIX

Locking the Falls

Financing the Improvements

E ven before the surveys and promotional tours were completed, Alfred Smith had quietly begun to negotiate for the purchase of the five existing upriver canal and lock operations. He reported to the directors of the Connecticut River Company in early 1825 that the upriver improvements could be purchased for $368,000. With this estimate in hand, the Company had a clearer vision of the overall costs of the enterprise. To the $368,000 for upriver acquisitions, they added the estimate of approximately $100,000 for constructing the canal at Enfield Falls. These figures did not include the expense of re-engineering the upriver canals and locks to permit the passage of uniform sized tow-vessels and barges nor the cost of dredging work to remove sandbars and other impediments. Although there were no solid estimates made for those expenses, it was clear to the directors that the authorized capital of $500,000 would be insufficient to achieve the corporate goals.

On May 5, 1825, the following memorial was presented to the Connecticut General Assembly:

> To the Honourable General Assembly of the State of Connecticut convened at Hartford on the first Wendesday of May, A. D. 1825, now in session. The Memorial of the Connecticut River Company by their President & Directors, respectfully showeth,
> That since obtaining the act of incorporation which your Honour-

able Body was pleased to grant to the Memorialists in May last, the Company has been organized and has raised a capital of 18,000 Dollars for defraying the expenses of procuring surveys of the river and estimates of the costs of the necessary improvements, and with the aid of skillful, practical Engineers, have made such survey of the Falls at Enfield as were requisite to ascertain the nature & extent of the works necessary to effect a safe easy and expeditious boat navigation; and that they have obtained accurate (though partial) estimates of the expenses of these works, from which it is evident that the amount will very greatly exceed the sum formerly calculated. Your memorialists also with the assistance of said engineers have partially explored Connecticut river from the north line of this State to Hanover in the State of New Hampshire; that they have examined the works of the existing Lock & Canal Companies, on said river, and have made such progress in their negotiations, for purchasing them, as to have ascertained that in order to obtain those works, to make the requisite alterations and improvements, and to construct such other works and make such other improvements on and near the river, as to accomplish in a satisfactory manner, the Great object of their incorporation, will require much larger Capital, than is allowed by their Charter. This object is of such vital importance to the vast and increasing population upon the borders of the Connecticut and its tributary streams, as justly to call into operation the greatest efforts of the community and to entitle the Memorialists to the liberal patronage of the Legislature. As an object of great national importance it has already attracted the attention of the General Government. To accomplish an object of such magnitude and public importance your memorialists respectfully pray your Honours to enlarge the Limits of their capital to a sum not exceeding One Million Five Hundred Thousand Dollars, and for that purpose to authorize the memorialists from time to time to open subscriptions for stock, to such an amount as may be necessary not exceeding the amount aforesaid. And the Memorialists further pray that for the purpose of enabling them to perform the various duties and services enjoyed by the said act of Incorporation, with greater economy and facility whenever they shall have expended One Hundred Thousand Dollars upon the objects designated in their charter they may be authorized to establish, in the City of Hartford an office of discount and deposit and to employ Banking operations, a portion of their Capital not exceeding what at any time shall have been previously expended upon the objects aforesaid and not exceeding in the whole the sum of Five Hundred Thousand Dollars, with the privileges of other incorporated Banks, under suitable limitations and restrictions. And that this Assembly would grant the memorialists such other relief in the premises as to them in their wisdom shall seem meet and proper and they as in duty bound will ever pray. Dated at

Three dollar note issued by the Connecticut River Banking Company.

Hartford the 5th day of May, 1825.

By order of the board,
Alfred Smith, President

A bank charter was a precious commodity during this period in the nation's history due in large part to a prevailing mistrust of banks in general and the resulting hesitancy of General Assemblies to grant charters for new banking ventures. The river improvement proponents were all individuals of wealth and local esteem, and their combined influence was impressive. The General Assembly was surprisingly quick to grant the Company's request. The Charter for the Connecticut River Banking Company was granted later the same month.[1]

The legislation revised the charter of the Connecticut River Company, allowing it to sell up to one million dollars in capital stock, doubling its original limit. The act then provided that as soon as the Connecticut River Company had expended two hundred thousand dollars on the objects of its incorporation, at least half of it on acquisitions and improvements north of the Connecticut border, the company could form the Connecticut River Banking Company. To raise capital for the banking company, the river company could assess each of its shareholders fifty dollars per share owned. The Connecticut River Banking Company was authorized to raise up to Five Hundred Thousand in capital stock, but never more than fifty percent of the amount of capital stock raised by The Connecticut River Company. The failure of a River Company

---

[1] *Resolves and Private Laws of the State of Connecticut*, p. 80.

shareholder to pay the assessment for the Banking Company stock would lead to a forced sale of one's shares. The failure of a Banking Company shareholder to pay any subsequent installments due on his or her River Company shares would result in a forfeiture of any dividends accruing on Banking Company shares.

The Connecticut River Banking Company's operations were limited to the trade of bills of exchange, gold or silver bullion, and secured lending transactions. Further, it was not permitted to issue notes totaling more than fifty percent of its capital stock and cash on deposit. Despite these limitations, the act concluded with one extremely beneficial provision:

> Whereas an improved navigation through the valley of Connecticut River is necessary to be made in different States, and will be of great public utility, the stock, property and income of said Connecticut River Company shall be forever exempt from taxation, and the stock of said banking company shall be exempt from taxation for four years from the next after rising of this Assembly.[2]

The tax exemption provided by the General Assembly made the stock of both companies a more favorable investment.

During the early part of the nineteenth century, the American banking industry was still in its infancy. As the nation's market economy grew, the need for banks and the services they offered grew with it. While the directors of the Connecticut River Company were eager to begin reaping the benefits of banking operations, they were delayed in forming the bank due to the requirement that they first spend at least one hundred thousand dollars on acquisitions and improvements north of Connecticut. They were powerless to do much of anything north of the border until they obtained legislative approval from the General Assemblies of the northern states. The efforts to obtain such approval were meeting with mixed results.

Legislative Approval in the Upriver States

The State of Vermont, settled principally by Connecticut transplants and considered by historians to be the progeny of Connecticut, was the first upriver state to grant the Connecticut River Company the right to oper-

---

[2] Ibid., p. 82.

ate within its borders. The enthusiasm generated at the convention promoting river improvement held in Windsor, Vermont in February 1825 carried over to the Vermont General Assembly. On November 9, 1825, the Vermont legislature authorized the Connecticut River Company to further its river improvement plans within its borders. It gave the company five years to complete acquisitions and improvements up to Brattleboro. Upon completion of improvements to Brattleboro, the company was given another four years to complete improvements to Barnet.[3]

New Hampshire confirmed the action of the state of Vermont shortly thereafter.[4]

Legislative approval in Massachusetts was a more difficult proposition. The battle between the Canallers and the Riverites was fiercely waged in the Massachusetts editorial pages and in the halls of the Massachusetts State House. Although the Hampshire and Hampden Canal, which would extend the Farmington Canal to Northampton, Massachusetts, was still years away from completion, the canal proponents were already advocating extending the canal to the northern border of Massachusetts. The sharp division among the competing interests led to a stand-off in the Massachusetts legislature.

In March of 1826, in an editorial noting the recent success of the Barnet's maiden voyage upriver, the editors of The Springfield Republican commented that the only obstruction met on the journey were at the several falls. They noted that the Connecticut River Company would make improvements to surmount those obstacles, "as soon as the legislature of Massachusetts shall have confirmed their charter."[5] Later that same month, *The Connecticut Courant* reported that the approval of the Massachusetts legislature had been obtained. A week later, the paper retracted the story, noting that the bill had passed only in the Senate. The House deferred the matter to its May 1826 session.[6]

When the May session opened, a competing petition to extend the Hampshire and Hampden Canal to the northern border of Massa-

---

[3] W. DeLoss Love, *The Navigation of the Connecticut River* (American Antiquarian Society, 1903).

[4] State of New-Hampshire. *An Act, to confirm an Act of the General Assembly of the State of Vermont, entitled "An Act to provide for improving the Navigation in the valley of Connecticut River."* (1826), Act,s p.30.

[5] *Springfield Republican,* December 13, 1826.

[6] *Connecticut Courant,* March of 1826.

chusetts had also been introduced. Both petitions were heard by committees but continued to the winter session. In the interim, the proponents of extending the canal began an intense lobbying effort in support of their cause. Petitions were circulated and the names of close to three thousand people supporting the canal extension to the exclusion of river improvement were submitted to the Massachusetts legislature. James Hillhouse, the chief proponent of the Farmington Canal and its Bay State counterpart, the Hampshire and Hampden Canal, brought in a high-profile figure to support the cause. New York governor and Erie Canal champion DeWitt Clinton was escorted by Hillhouse on a tour of the completed portions of the canal and those under construction. Additionally, they toured the Connecticut River valley north through Massachusetts, Vermont and New Hampshire. The terminus of the surveying tour was very publicly, and not coincidentally, announced to be Barnet, Vermont, widely known as the "promised land" of Connecticut River Company aspirations.[7]

Fearing the advantage that a favorable report from Clinton would bestow upon the canal cause, *The Boston Journal* published a scathing editorial regarding his visit. Bostonians did not favor the canal plan as it would make New Haven a port of entry that could rival Boston Harbor. Boston was already exploring the possibility of a canal from Boston harbor to Troy, New York, a direct link from Boston to the Erie Canal and the Great Lakes. The editorial, which was reprinted in the *Connecticut Courant* on June 4, 1827, provides:

> Gov. Clinton is making a progress through the western part of Massachusetts to Vermont, for the purpose, it is said, of viewing the proposed route of the Hampshire and Hampden Canal. Although Mr. Clinton is a man whom the citizens of Massachusetts must always feel honored and happy in receiving as a visitor and a guest; still the occasion of the present visit appears to us singular. Does he intend to express his opinion of the practicability, and expediency of the canal in comparison with its rival enterprise, the Connecticut River Navigation? This we think cannot be, for he must know that the comparative merit of these two projects are a subject of legislative inquiry, and are to receive a legislative decision in our General Court. It would, as it appears to us, be a libel on the Governor of New York to say the he intends in any way to interfere in a local question of much delicacy and importance, a question which has aroused large and respectable portions of the Commonwealth to keen,

---

[7] David Hosack, *Memoir of De Witt Clinton With an Appendix, Containing Numerous Documents, Illustrative of the Principal Events of His Life* (1829).

and almost angry controversy, a question upon whose decision property to an immense amount is depending! It can scarcely be presumed that Mr. Clinton has any political views in undertaking this tour any more than he had in undertaking a former one to the west. There is, however, a manifest difference between the two cases. In relation to the Ohio Canals, there were no parties, and there was, we believe, an invitation from the Executive, but we are not aware that there has been any invitation from the Executive of Massachusetts. Indeed it is utterly impossible from the nature of the case, that our Governor could have invited the accomplished chief Magistrate of N. Y. to meddle with matters in which Mr. Lincoln himself has no right to interfere. It may be added also that the city of New York is supposed to have a direct interest in the success of the canal, an interest adverse to the metropolis.[8]

Clinton was apparently aware that he had stepped into a hornet's nest of controversy. In his subsequent letter to the Hampshire and Hampden Canal Company, he stated,

Having no other object in view than the interest of internal improvement, I should greatly regret if my visit was misconstrued into an intrusive intermeddling with the concerns of other states, or an officious interference with existing controversies.[9]

He was careful in the balance of his letter not to touch upon the relative merits of canal construction as opposed to river improvement. Nevertheless, he waxed poetically about the benefits of a canal, saying,

Towns and villages would spring up in every direction, and the wilderness and the solitary place will become glad, and the desert rejoice and blossom as the rose.[10]

The majority of Massachusetts legislators were unimpressed by Clinton's prose, or perhaps they were too busy trying to figure out where in New England one might find any desert, let alone a rejoicing one. Neither the canal extension bill nor the river improvement bill was acted upon in the winter session of 1827-1828. The sides were as divided as ever, and division perpetuated the impasse.

---

[8] *Boston Journal*, June 4, 1827, Reprinted in *The Springfield Weekly Republican*, June 6, 1827.

[9] Hosack, *Memoir of DeWitt Clinton, p. 222.*

[10] Ibid.

A Further Impediment - The Enfield Bridge Company Strikes Again

The Directors of the Connecticut River Company knew that they could wait no longer for the Massachusetts legislature to act. The Company needed to move the river improvement project forward or risk losing the confidence of its investors. The Enfield Falls were entirely within the boundaries of the State of Connecticut. No authority from Massachusetts was needed to improve navigation on that section of the river. A lateral canal circumventing the Enfield Falls was the first, and most instrumental step, to implementing the river improvement scheme.

By the spring of 1827, engineering plans were finalized, land acquisition began, and a work force was assembled by the Connecticut River Company. No sooner had the project commenced than it was brought to a sudden halt by a familiar and notorious foe. The Enfield Bridge Company founded by John Reynolds had in 1798 been awarded the first franchise to lock the falls. This franchise had been awarded in conjunction with the Company's franchise to build a toll bridge between Enfield and Suffield. For years, Reynolds engaged in a game of cat and mouse with the Connecticut General Assembly as he sought to profit from the benefits of the bridge without bearing what he considered to be the burden of the locks. In this effort he was amazingly successful. By November of 1808, he had completed the bridge, had begun to collect tolls, and had been discharged from the obligation to build the locks around Mad Tom and Surf Bar, the principal rapids of the Enfield Falls.

John Reynolds died on July 4, 1812 at the age of 74. Rufus Granger of Suffield acquired a controlling interest in the Company.

When the Middlesex Canal hero, John L. Sullivan spearheaded a second attempt to bypass the Enfield Falls in 1818, it was The Enfield Bridge Company that again stymied the effort, this time by using their original franchise rights to secure a controlling share of Sullivan's enterprise. True to form, it was the Enfield Bridge Company which again sought to block the latest effort by the Connecticut River Company.

In June of 1827, The Company for Erecting and Supporting a Toll Bridge with Locks from Enfield to Suffield filed a suit in the Connecticut Superior Court seeking an injunction preventing the Connecticut River Company from building a canal which, it claimed, would be a direct interference with its own franchise rights. In the ensuing years since the Bridge Company last thwarted efforts to lock the falls, a controlling interest in the Bridge Company had been acquired by prominent Suffield resident Rufus

Granger. Granger was no less devious than Reynolds and he was appointed as agent by the Bridge Company to prosecute the case on behalf of the Company.

The case was heard by Judge Titus Hosmer during the February, 1828 term of the Court. In support of its petition, the Bridge Company introduced into evidence an 1808 certificate from the Connecticut General Assembly accepting the completed bridge and authorizing the company to begin collecting tolls. Additionally, the Company introduced a deed dated June 20, 1827, to a small piece of land opposite Mad Tom and Surf Bars. The deed was evidence, the Company argued, of its intent to shortly commence construction of locks.

The Connecticut River Company countered with several arguments, summarized as follows:

1. That the Bridge Company never provided the bond that was required by its charter before building its bridge, thereby rendering its charter void.

2. That the Bridge Company, having already suffered the destruction of the original bridge by floodwaters, had offered to sell the remaining piers of the bridge, thereby indicating its intent to forfeit its franchise.

3. That any locks contemplated by the Bridge Company would be of no public benefit.

4. That the bridge in question had not been in existence for more than ten years.

5. That Rufus Granger had used false and fraudulent representations to the Bridge Company stockholders to acquire his shares.

6. That neither the locks authorized by the General Assembly in 1798, nor any subsequently approved shore channels, had been built and, if built, would not benefit navigation.

The hearing on the injunction was held in the Connecticut Superior Court in Hartford before the Honorable Steven Titus Hosmer. Steven Titus Hosmer was the son of Titus Hosmer, a member of the Continental Congress. He was a lifelong Middletown, Connecticut resident. Hosmer graduated from Yale in 1782 and began his practice of law in Middletown in 1785. After the establishment of an independent judiciary in Connecticut, it was not uncommon for a Superior Court judge to simultaneously sit on the Supreme Court of Errors, the State's highest Court. Hosmer served as the chief justice of the Supreme Court of Errors from 1815 to 1834. Middletown was a rival port to Hartford, and had been losing ground to Hartford economically for many years. The Farmington Canal plan represented as much a threat to Middletown as to Hartford and the upriver improvements would have benefited Middletown as much as Hartford. As a Middletown native, Hosmer was intimately familiar with the maritime affairs of Connecticut.

After hearing the case in the Superior Court, Judge Hosmer specifically and methodically rejected each of the arguments advanced by the Connecticut River Company. It appeared certain that The Enfield Bridge Company would once again derail an effort to bypass the Enfield Falls. It is unlikely that the company had any objection to the concept of locking the falls. It is more likely that the proprietors were simply opportunists who recognized that while the franchise seemed more of a burden than a benefit to them, it might have more appeal to others. While the proprietors of The Enfield Bridge Company never believed that actually constructing the locks could lead to a profit, they understood that they might profit by the sale of the franchise rights to others who were more optimistic. After the hearing, the Enfield Bridge Company appeared to be on the brink of success. If the injunction issued, the Connecticut River Company would be forced to purchase the franchise rights or abandon the project. Perhaps due to his simultaneous service on the Supreme Court or perhaps sensitive to the impact his decision would have on the general public, Hosmer did not issue an injunction. Instead, he referred the matter to the Supreme Court for advice, particularly as to whether his determinations in rejecting each of the Connecticut River Company's defenses were correct.

The Connecticut River Company had dodged the bullet, but its directors could not be particularly hopeful for a favorable result before the Supreme Court, for among the other justices on the Supreme Court was the influential New Haven jurist David Daggett. Daggett was one of the founders of Yale Law School. In such capacity, he was closely aligned with Yale treasurer, James Hilhouse, who was the principal proponent of the Farmington Canal. Daggett's New Haven ties were so thorough that, later in the same year that he would hear the Connecticut River Company case, he was elected Mayor of the city. The other justices who would hear the case were John Thompson Peters, who was educated at Yale and practiced law in his native Hebron, Connecticut, before being appointed to the bench and James Lanman, who hailed from Norwich, Connecticut and also graduated from Yale.

The written opinion in the case was authored by Justice Hosmer with a concurrence by Justice Daggett. Not surprisingly, Justice Hosmer reiterated his rejection of each and every one of the Connecticut River Company's defenses. Daggett did the same in his concurrence. Amazingly, despite the Connecticut River Company's abject failure to persuade the justices that there was merit to any one of its arguments, the injunction sought by the Bridge Company did not issue. Hosmer ruled:

> The public have a deep interest in the commodious navigation of the Connecticut River, and it is peculiarly inequitable, that the rights of a community should be sacrificed, to insure the franchise of the plaintiffs from all possible damage, while they are in actual enjoyment of it, and have taken no measures to pay the price of their charter. This, however, is not all. Twenty years have elapsed, and nothing has been done, by the plaintiffs, to benefit the navigation of the Connecticut River. For this extraordinary degree of laches no apology is derived from the omission of an order by the General Assembly. No application has been made to them, by the plaintiffs, to turn their attention to this subject. Nor does it appear, that any effectual measures have been taken, or are even contemplated, to bring this lethargy to a termination. A small piece of ground has been purchased, and this is all; except the declarations now made of an invisible intention to do what might and ought to have been done long since.
>
> For the imposition of the court, I discern no justifiable ground, and advise that the plaintiff's bill be not granted.

Daggett was no less critical of the Enfield Bridge Company in his concurrence, ruling,

> …to grant the injunction prayed for, will be to permit the Enfield Bridge Company to lie by, hold this franchise, and prevent the legislature from authorizing any other company to do what in 1809 was expected, by the state and by the plaintiffs, viz. locking both the upper and lower falls.[11]

The inaction and insincerity of Reynolds and his successors had finally caught up with them. Without denying the Bridge Company's franchise rights, the Supreme Court determined that it had sat on its rights for so long, that the public interest demanded that those rights not stand in the way of a project of such public importance. Even the New Havener, David Daggett could not deny that improving the navigation

---

[11] *The Company for Erecting and Supporting a Toll Bridge with Locks from Enfield to Suffield v. The Connecticut River Company* 7 Conn.28 (1828).

of the Connecticut River was a public good that could not be overcome forever by the economic interests of a handful of selfish men. The final impediment to bypassing the Enfield Falls had finally been cleared.

CHAPER SEVEN

Building the Canal

The Chief Engineer – Canvass White

In February 1827, the Connecticut River Company offered to engage Canvass White as Chief Engineer. The offer letter expressed the desire of the Directors of the Company to "construct a canal by Enfield Falls, the present year."[1] The proposal also requested White to make an examination of the river to further develop the plans for a general improvement of navigation between Hartford, Connecticut and Barnet, Vermont.[2]

The written proposal recites the understanding of the Directors that White was willing to devote four months of his time during 1827 to the project, that he would pay his own travel expens-

---

[1] Undated Resolution, Canvass White Papers, Rare and Manuscript Collection, Carl A. Kroch Library, Cornell University Library.

[2] Ibid.

es, and that he would accept for these services the sum of $2000.[3]

On March 12, 1827, White responded:

> Gentlemen,
>
> I do hereby engage to take charge of the contemplated improvements at Enfield Falls as principal Engineer for one year and to arrange and digest plans for the improvement and navigation of the River as may be directed by the Company and to devote about four months of the year to the above objects for the sum of two thousand dollars payable semiannually including expenses while engaged on the works at Enfield Falls. The Company to pay my travelling and contingent expenses while engaged on other parts of the line and also to furnish an Assistant Engineer and other necessary assistants of my selection. I am to have no charge of making contracts other than giving an opinion as to prices when requested and also to have no charge of the payments of money other than that connected with the expenses of the Engineering departments.[4]

There are inconsistencies between the offer and the acceptance that would lead to some conflict down the road, but the agreement was specific enough for White to commence work on the project.

An undated document among the Canvass White papers, located in the same folder as the offer and acceptance letters, is a list of four questions.[5] The list confirms that despite White having proposed several potential designs for the canal in 1824, the precise design of the canal had not yet been determined as of that date White was engaged. The questions were as follows:

> What is the best mode of making a canal by Enfield Falls on the east side and on the west side of the river?
>
> What will be the probable difference of expense?
>
> On which plan will the works be least exposed to damage & the navigation to be interrupted?

---

[3] Ibid.

[4] Letter to President and Directors of the Connecticut River Company, dated March 12, 1827. Canvass White Papers, Rare and Manuscript Collection, Carl A. Kroch Library, Cornell University Library.

[5] Undated Document. Canvass White Papers, Rare and Manuscript Collection, Carl A. Kroch Library, Cornell University Library.

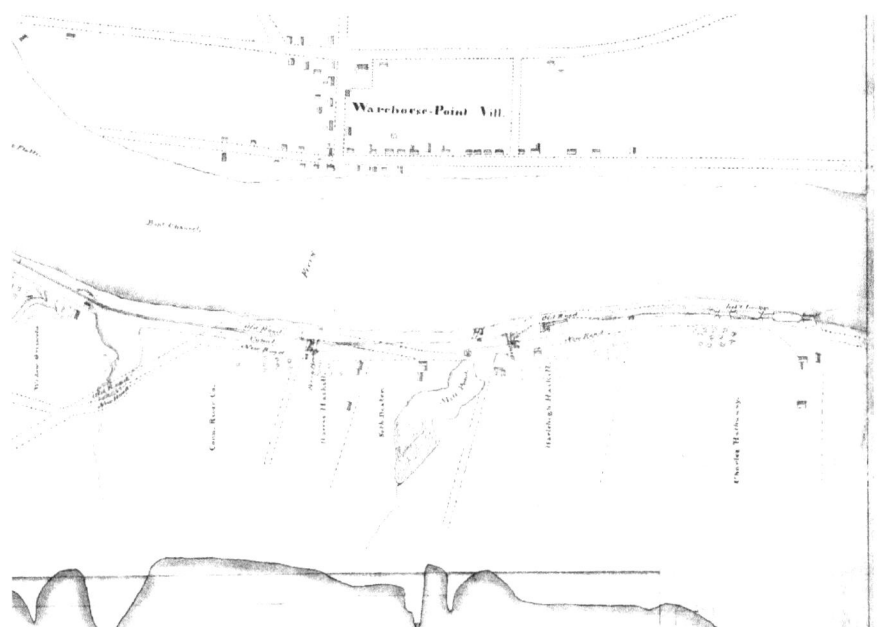

Section of Canvas White's engineered drawing of the Windsor Locks Canal at and above lower locks.

> Which side is in your opinion best taking into view cost of construction, damage to private property, safety of the works, and convenience to the navigation?

A May 11, 1827, report from White is responsive to the list of questions as it begins, "In reply to the several questions put into my hand by you on my departure from Hartford…" He then restates the first question, "What is the best mode of making a canal by Enfield Falls on the east Side and the west side of the river?" White then threw the Company an apparent curveball by explaining in the report that he has "changed his opinion very much as to the contemplated improvements" and that the best plan has "not heretofore been laid before the board." The report goes into detail explaining the new proposal, which included building two dams, one connecting the northern head of Kings Island to the east bank of the river, and one connecting the southern foot of Kings Island to the west bank of the river. White suggested that this design would eliminate the need for the upper shore channel on the west bank of the river as the water level would be raised generally over all the upper rapids. The canal below Kings Island would proceed overland as originally suggested. The dam running east of Kings Island would be designed with sluices and gates for the passage of fish and boats navigating down the river if "they do not choose to pass through the canal."[6]

To the board members of the Connecticut River Company, this late shift in thinking may have been reminiscent of the maneuverings of John Reynolds and Company a decade earlier. White was also somewhat elusive as to how these proposed modifications would impact the cost estimates. The Company may have doubted that its corporate charter would even allow a design that effectively dammed the river entirely. The Company wasted no time in rejecting the proposed design change. White must have capitulated quickly, for just five weeks later he released a detailed map depicting the project exactly as it would be built. The map is intriguing in its detail, including all canal features, abutting properties with name of owner and location of buildings, tributary streams, hazards to navigation in the river including the identification of known rapids by name and the piers of the former Enfield Bridge. The map's detail is so accurate that it correctly identifies the ownership of properties acquired by the Connecticut River Company just two weeks before the map was issued. Along the bottom of the map is an elevation profile. In the upper right corner of the map is a lengthy text box, which has been painstakingly transcribed by the late Suffield Town Historian Lester Smith:

> Map of part of Connecticut River, including Enfield Falls. Also, of the Canal at said Falls as laid out, under the direction of the Commissioners, pursuant to an Act entitled an Act to Incorporate the Connecticut River Company. Also, the new Road or public highway as altered with the consent of the Commissioners, by reason that the Canal could not be judiciously laid out without interfering with the present road.
>
> The works begin at the head of Enfield Falls, at Matthew Thompson's land, by a dam extending from the East shore towards the West shore raised to a level one foot higher than the surface of the flat rocky bed of the river at and above Dollar Bar, leaving open a space not less than fifty feet in extent and which shall include the ordinary boat channel down the river. The deeper parts of the Boat channel to be partly filled up, or raised but not reducing the depth therein to less than two and a half feet at low water, nor less than the shoalest places in the other parts of Enfield Falls, in the boat Channel.
>
> From the West end of the dam aforesaid a wing dam to pass obliquely, by Widow Kendall's land toward the West bank on Samuel Norris' land,

---

[6] Document dated May 11, 1827. Canvass White Papers, Rare and Manuscript Collection, Carl A. Kroch Library, Cornell University Library.

thence a pier in the bed of the river on or against said Norris' land to be there joined with an embankment opposite guard lock which is to be made near the shore of' said Norris' land and to be connected with the pier & embankment by a breast wall with sluices for taking water into the Canal. From the south end of the pier an embankment of earth and stone and timber, in the bed of the river to extend along and by the land of said Norris, of Levi Pease, David Pease, Seymour King, Diadama Alden, James Ives, Dan King, Abiel King, Isaac King, Tabitha Simonds, Levi Pease, Dan King Jr., Rufus Granger & Henry Pease, thence passing from the bed of the river the Canal will cross Stoney Brook by an aqueduct with stone piers & abutments, and wooden trunk, then on Henry Pease' s land South 23° 30' West two chains, S. 12° W. one chain, S. 3 ° W. two chains, S. two 90/100 chains then on John Moran 's land S. four 10/100 chains, S. 5° 30' E one chain then on Allen Loomis' land S. 5° 30'E one 80/100 chains S.14° E 3 chains, S. 24° E three chains, S 32° E. three chains, S. 42° E three chains to the river, then in the bed of the river by an embankment as before on & by land of Allen Loomis, Nathaniel Parmelee & his wife, Elisha Sperry, Rufus Granger, Enos Moron, Jerusha Moron & Widow & heirs of Darius Parmelee, to land of Corning Fish and wife, at the South end of the bluff where the Canal and embankment again leave the river bed, and pass along on land of said Fish and wife N. 85° 30' West three chains S 85° 15' W. two 80/100 chains S. 776°W three chains, S. 55° 30' W. three chains, S. 38°W. three 10/100 chains, then on John Moron's lands. 35° 30' W. twelve 81/100 chains, S. 45° 30' W three chains, S. 54° 45' W. four chains S. 43° 30' West two chains, S 30° W. two chains S 28° 30' W. nine chains, then on land of the Widow or heirs of Erastus Griswold S. 28° 30' W. 12 chains, on the last lot a waste Weir, then on land of the Connecticut River Company S 22° 45' W two 67/100 chains, S. 13° W. two 94/100 chains, S 8° West twenty four 68/100 chains S. 4° 30 W. one chain, then on land of Harris Haskel S. 4° 30' W nine chains, then on land of Seth Dexter S 4° 30' W seven chains S. 1° W. four chains, then on land of Herlehigh Haskell S 4° E. four chains S 10° E. three chains S 11° 30' E fourteen chains then on land of Charles Hathaway S 11° 30' E two chains S. 6° E three chains S. 2° W. five chains S. 3° W. fifteen chain's thence in the River as far down as said Hathaway's Wharf. Three stone lift locks to be placed within the last twenty chains. The embankment to form the Canal is coloured on this Map light red. - the land flowed or covered with water by the Canal is on the Westerly side of the embankment, and coloured blue - the new or altered road lies between Charles Hathaway's South line and the fork made by meeting of the road to Suffield and the river road to Springfield, is laid out on land of Harris Haskell & Herlehigh Haskell, Connecticut River Company, Seth Dexter and said Hathaway, is shaded brown and marked new road on this Map. And said Connecticut River Company by their Agents, superintendants or engineers may enter into and take posession of' all the said lands, waters and streams on this map necessary for making said dams, pier, embankments &

the prosecution of the improvements intended by said Act.

> Windsor. June 20th. 1827. certified by
> Asa Willey
> Martin Welles
> John S. Peters
> [signatures]
> Commissioners under oath[7]

Once the general layout and design had been agreed upon, White set out to produce specifications for each aspect of the work. On September 2, 1827, he submitted his work to the Connecticut River Company. Generally, he explained that the canal was to be seventy feet wide at the surface, fifty-two feet wide on the bottom, and six feet deep. He calculated that the area of the canal at any single cross section would be 366 feet.[8]

White submitted individual specifications sheets for several aspects of the work as he agreed to do when he accepted the position of Chief Engineer. These included specification sheets for the sluices to supply water to the canal, the breast wall in which the sluices are secured *(now, in combination with the sluices, often called the "head gate")*, the abutments to support the breast wall, the pier to connect the guard lock to the breast wall, the pier connecting the guard lock and breast wall to the wing dam at the head of the falls, and the aqueduct over Stony Brook. Some of the specification sheets include illustrative drawings. Some, as well, include calculations as to the number of "sticks" of lumber that would be required for each component.[9]

The sheets submitted did not include specifications for the construction of the guard lock on the north end of the canal and the three lift locks on the south end of the canal. The reason for this omission is that no determination had yet been made as to the manner of construction of the locks. Instead, White included a letter to Alfred Smith noting he had "reflected a good deal on the construction of wooden locks as

---

[7] Map of Part of the Connecticut River Including Enfield Falls, Cartographer unknown. Prepared under the supervision of Canvass White, Chief Engineer of the Enfield Falls Canal, Watkinson Library, Trinity College, Hartford, 1827, Transcribed by Lester Smith.

[8] Undated Document. Canvass White Papers, Rare and Manuscript Collection, Carl A. Kroch Library, Cornell University Library.

[9] Ibid.

proposed for the Enfield Falls Canal." He then opined, "locks built with rough stone laid in cement lined with plank will be decidedly preferable to timber locks with stone heads." He explained that the lower two locks would be subject to occasional flooding and the best way to secure them is to build them fully of stone and protect them with embankments of earth. White estimated that 300 perch of stone would be required for timber frame locks and that an additional 1000 perch of stone would be required for locks fully constructed of stone. He calculated that the cost differential would be $2000 per lock, but he reminded Smith that only four locks were called for in the design so the total cost differential would be modest compared to the quality gained. He urged Smith to convince the board to choose locks constructed completely of stone.[10]

In the same letter, White explained that he had further considered the manner of passing Stony Brook and concluded that the best design would be "an aqueduct of 50 feet wide and 6 feet deep to be constructed of wood and stone, so as to be easily repaired – instead of a dam and guard gates and a culvert for the paper mill as formerly proposed." He included a specification sheet for the aqueduct as proposed among the other sheets submitted with the letter.[11]

The Assistant Engineer – Edwin A. Douglass

In his response to the Connecticut River Company's offer to engage him, Canvass White added the condition that the Company pay for an Assistant Engineer of White's choosing. White chose Edwin A. Douglass *(sometimes spelled Douglas)* as his Assistant Engineer. Douglass was born on March 3, 1804, in Stephentown, Rensselaer County, New York. He was among the core of young civil engineers who acquired their professional skills working on the Erie Canal. He undoubtedly became acquainted with Canvass White during that time and the two men maintained a close professional connection until White's death in 1834. Douglass succeeded White as the Chief Engineer on several unfinished projects, including the Lehigh Canal.

---

[10] Correspondence to Alfred Smith dated September 3, 1827, Canvass White Papers, Rare and Manuscript Collection, Carl A. Kroch Library, Cornell University Library.

[11] Ibid.

The Lehigh Coal and Navigation Company constructed the Lehigh Canal from Mauch Chunk *(now Jim Thorpe)*, Pennsylvania, to Easton, Pennsylvania, to provide a means of transporting anthracite coal to Philadelphia. Work on the canal commenced during the summer of 1827 under the direction of Canvass White. As Edwin Douglass was engaged on the Enfield Falls Canal at that time, he did not immediately join White in that endeavor, but assisted him on other projects once he was available.

In 1835, the Lehigh Coal and Navigation Company decided to extend its canal 26 miles north from Mauch Chunk to White Haven. Edwin A. Douglass was named Chief Engineer of this project, which included 29 locks and 20 dams. The work was completed in 1838. The upper division of the canal was in operation until 1862.

From 1843 until his death in 1859, Edwin Douglass was the superintendent and engineer of all the Lehigh Coal and Navigation Company works. In this capacity, he designed, prepared estimates for, and oversaw the improvements to the famous "Switch-Back" gravity railroad that transported anthracite coal from the elevated coal fields to the Lehigh Canal. In addition to its coal carrying function, the Switch-Back Railroad became one of the nation's most popular tourist attractions, and the income derived from charging thrill-seekers to ride in the cars as it careened down the slope of the mountain greatly supplemented the company's income.

Chief Engineer Canvass White also periodically sent instructions to Assistant Engineer Douglass about items of particular concern on the Enfield Falls Canal. For example, on August 8, 1827, White wrote to Douglass: "I wish you to pay particular attention to the slope wall along the river embankment and be rigid in enforcing the directions given."[12]

When White came to Enfield Falls, he and Douglass walked the work site to make sure Douglass understood exactly how White wanted things done. While the men are not identified, it appears that local historian Jabez Haskell Hayden observed one of these on-site meetings. He wrote:

> Previous to the commencement of work of digging the canal at Windsor Locks some gentlemen in charge with their engineer came on the ground to decide what should be the slope, the angle of the inside of the canal necessary to prevent the new made bank from sliding in when the water should

---

[12] Canvass White to Edwin Douglass, August 8, 1827. Canvass White Papers, Rare and Manuscript Collection, Carl A. Kroch Library, Cornell University Library.

be let in. I was here on a visit at the time and followed on to see what-ever pertained to the proposed canal. They went over the river bank and selected a spot where the Dexter paper mill now stands, then they graded off the bank until they obtained what they thought a suitable pitch or grade. When the man of the shovel raised a laugh by exclaiming "if they will make it as good as that I'll compounce it good." The engineer made a note of the angle of the slope decided upon for future reference.[13]

While supervising the construction of the Enfield Falls Canal, Edwin Douglass was introduced to Harriet C. Dexter, daughter of Seth Dexter and sister of papermaker Charles Haskell Dexter. A romantic interest soon developed. Despite the distance between the two, after the Enfield Falls Canal was completed and Douglass joined Canvass White to assist him on other projects, the romance persisted. Harriet Dexter and Edwin Douglass were married on February 6, 1834, in Windsor. Harriet then joined Douglass in Mauch Chunk, Pennsylvania where they resided thereafter. Harriet Dexter Douglass died in 1846 just a month after giving birth to their fourth child, Charles Edwin Douglass. She was buried in Grove Cemetery, Windsor Locks.

Due to the family connection, Edwin A. Douglass was encouraged to join the Dexter family paper-making business. In 1847, he joined Charles Haskell Dexter in incorporating C. H. Dexter and Company, the longest surviving mill on the Windsor Locks Canal. He soon left the enterprise, as the sedentary corporate life did not suit a man who had spent his entire adult life outside on canal projects. Edwin Douglass died on December 23, 1850. His remains were buried next to Harriet in Grove Cemetery.

The Contractors

The Connecticut River Company engaged contractors to complete the Enfield Falls Canal. This was the system that developed during the construction of the Erie Canal and the system with which Canvass White was familiar. Separate contracts were let out for eleven sections of approximately one-half mile each and for the locks and aqueduct. The following notice appeared on or about May 13, 1827, in newspapers throughout Connecticut, Massachusetts, New Hampshire, and Vermont:

---

[13] Hayden, pp. 26 -37. *op. sit.*

NOTICE

On the 21st of May instant at Enfield, also in the Office of the Connecticut River Company in Hartford, proposals will be received for constructing a canal by Enfield Falls. The works will be divided into convenient sections, and will embrace embankments in the river with Sloping Stone Walls, Excavation, a Wing Dam, and Locks. Canvass White, Esq., Engineer of the Company, will attend at the Falls to settle and explain the location and manner of construction of the works. By order,
A. SMITH, PRESIDENT

Office of the Conn. River Company
May 12, 1827

From the correspondence retained in the Canvass White papers, as well as other documents, the identities of some of the contractors can be determined.

The Masonry Contractor – Col. Mason Barker

After consulting with Alfred Smith, and considering the time remaining in the 1827 building season, Canvass White and Edwin Douglass decided that they would not commence construction of the stone structures of the canal until 1828. The plan was to have all the stone and other building materials delivered to the site so that the work could commence as early as weather and water levels would permit. For this reason, the Connecticut River Company did not rush to engage a masonry contractor. A letter from Edwin Douglass to Canvass White on August 30, 1827, notes that Alfred Smith and Connecticut River Company Directors Joseph Pratt and Samuel Belcher were on site negotiating with "the Browns" to take the contract for the wing dam and locks. The Browns were not engaged for this work, and it is likely that Canvass White had already decided by that time who he wanted to be his masonry contractor.

The masonry contractor who was ultimately engaged to build the four locks and the aqueduct was Colonel Mason Barker. Barker was from Schuyler, Herkimer County, New York. As a young man, during the War of 1812, Canvass White joined a New York militia and participated in the assault on Fort Erie. White was wounded during that conflict. The fact that Mason Barker maintained the title Colonel suggests military service as well. It is likely White and Barker served together. Canvass

White and Mason Barker surely worked together during the construction of the Erie Canal. As a mason, Barker was a frequent customer of White's hydraulic cement manufactory, as the White brothers were the sole domestic providers of that essential canal building compound.

On January 14, 1828, Barker wrote to White from Schuyler, New York:

> Dear Sir,
>
> I am anxious to see you as I have not had the pleasure of seeing you since I closed the contract with the Connecticut River Company for the 4 Locks and Aqueduct…I saw your brother at Milford. I had some conversation with him about furnishing the cement necessary for the works at Connecticut…If I remember, you concluded 100 barrels would be a sufficient quantity of the NY cement for the four Locks.[14]

The letter is formal, as most letters of that time tended to be, but there is also a tone of familiarity that suggests a longstanding business and personal relationship. Barker concluded the letter by recommending a childhood friend for the position of gate-tender at the Cohoes Bridge Company, another of White's business holdings.

On March 14, 1828 Mason Barker wrote to White that he entered into the contract with the Connecticut River Company "after you left Hartford last Oct." He also notes in the letter that he has a contract to perform work on the Lehigh Canal, a simultaneous project for which White was also serving as the Chief Engineer. This note confirms that White and Barker had a professional connection that extended to at least two of White's canal projects. The purpose of the letter was to provide White with an update on the progress of his work on the Enfield Falls Canal. Barker wrote:

> I procured a bar of iron and sounded the two upper lock pitts, of which the first is excavated out to the bottom. The first lock will require driving piles not exceeding 7 or 8 feet long. At that depth we found hard pan or rock. Lock No 2 – the hard pan or rock, and I think the latter, is not more than 2 to 2 ½ feet below bottom, which may be taken out down to the rock and filled in with good hard gravel in place of the sand up to bottom. May answer for a foundation to lay the timber on. Yet whatever should be

---

[14] Mason Barker to Canvass White, January 14, 1828. Canvass White Papers, Rare and Manuscript Collection, Carl A. Kroch Library, Cornell University Library.

your directions in the premises shall be strictly adhered to on my part -the high water at present may prevent me from making any certain conclusions respecting the foundation for the lower lock. You mention something in your note of 4th instant about the timber for the Locks on the Enfield Falls Canal. I suspect you must allude to that handsome lot of lock gate timber which we saw on the bank of the river in front of Mr. Lester's Inn last fall and now since I have not had the pleasure of an interview with you on that if that be it and on many other subjects I shall expect some communications in the premises either before you leave Albany or before you should Hartford and whatever should be your instructions and Directions relative the construction of the Locks and Aqueduct on the Enfield Falls Canal will be swiftly complied to on my part and I should be very happy to hear something from you concerning my Locks on the Lehigh Canal.[15]

While the plan was to have all material in place so that Barker could commence his work early in the Spring of 1828, Edwin Douglass noted in a letter to White on March 8, 1828 that, "The man that engaged to get the stone for the Locks has also failed and contracts made with others persons." Barker also noted in his March 14, 1828, letter that, "I have made such arrangements for stone and cement as I think will enable me commence the work by the first of May next." In fact, Col. Barker began his work sooner than that. In his April 24, 1828 letter to White, Edwin Douglass reported on Barker's progress:

Col. Barker left here previous to my receiving your letter. He commenced his work pretty spiritedly. The piles in the upper Lock is about 2/3 driven. The lower side they drive from 9 to 11 feet but on the upper side from 7 to 8 feet & pretty solid. They expect to commence laying timber next week. Will it be necessary to treenail the timber on the piles? I think the second pit will require piling it appears to be quite hard on the top of the gravel but after penetrating about a foot into it, it grows softer.[16]

The available evidence leads to the impression that Col. Mason Barker was a skilled and efficient masonry contractor, and it appears the Connecticut River Company was fortunate to have engaged him to work on the Enfield Falls Canal.

---

[15] Mason Barker to Canvass White, March 14, 1828. Canvass White Papers, Rare and Manuscript Collection, Carl A. Kroch Library, Cornell University Library.

[16] Edwin Douglass to Canvass White, April 24, 1828. Canvass White Papers, Rare and Manuscript Collection, Carl A. Kroch Library, Cornell University Library.

Mr. Lowell and Mr. Loomis - Sections Two and Three

The sections to be let out to contractors were numbered one through eleven running north to south. Section One included the guard lock, the head gate with sluices, and the wing dam. The sluices were required to allow water into the canal and precisely control its level. This function was important for the canal's navigational purposes, but it was especially important because the Connecticut River Company planned to sell water to produce hydropower to mill operators it hoped would spring up along the canal bank. The water would enter the mill races through apertures in the east wall of the canal. The amount of water drawn off for each mill, and therefore the price paid for the water privilege, was controlled by the number of apertures and the width and depth of the apertures. One constant was required, the water level in the canal needed to be equal to the top of each aperture. If the water level dropped below the top of the aperture, the mill owner would not receive the quantity of water he was paying for. Hourly readings were made, and the sluices could be cranked open or closed as needed to maintain this constant water level.[17] Since Section One primarily involved masonry, this Section was part of the contract with Col. Mason Barker. It is possible that Barker subcontracted out some of the work to others, such as the wing dam, as this component was primarily a wooden structure filled with loose stone and gravel.

North of the current railroad bridge, the canal was not excavated from existing dry land. The canal was formed by erecting embankments in the river running parallel to the west bank of the river. The specifications for the embankment utilized timber crib wharf construction. Wooden timbers were used to build giant cribs set in the desired place. The cribs were then filled with loose stone, gravel, and earth to permanently set them in place. The embankment was created by layering filled cribs upon filled cribs. A cross section of the embankment would appear as an isosceles trapezoid. On the Suffield portion of the Enfield Falls Canal, the sheer rock walls that now form the west bank of the canal were once the sloped walls on the west bank of the river. Material was blasted or otherwise removed from the sloped west riverbank to form perpen-

---
[17] The C. H. Dexter Company Records, Series VIII: Affiliated Businesses (1848-1936), Subseries A: Connecticut River Company (1848-1926), University of Connecticut. Archives & Special Collections, UConn Library, Dodd Center for Human Rights.

dicular rock walls. Wheelbarrows were then used to cart the excavated material up narrow plank runs to dump into and fill the timber cribs. The cribs were then covered with earth. Cut stone was then laid along the angled walls of the embankment. Lining the embankments with cut stone was a feature unique to the Enfield Falls Canal. That enhancement was necessary because the planners of the canal were among the first to anticipate using steam powered tugs to transport barges loaded with goods through the canal. There was concern that the wash from the steam tugs would erode the canal walls, so stone was added to the design to prevent erosion of the walls. Lastly, a tow path *(which now forms the Windsor Locks Canal Trail)* was placed on top of this embankment so that vessels with no mechanical power could be towed by horses and mules walking along the towpath.

Jabez Haskell Hayden discussed observing the construction of the embankments many years later. He wrote:

> Around the bluff at the present railroad bridge the river bank was abrupt, rocky and high and it was necessary for some distance at that point and above to lay the outer bank (the tow-path) of the canal in the bed of the river. Runs were made for wheelbarrows extending from the top of the bluff and out over the proposed tow-path, and the earth wheeled out and dumped on to framework prepared to receive it below… They dug and removed the earth with wheelbarrows on plank runs of the width of a single plank, and when near the bottom of the excavation it required a run of steep grade, and strong muscles, to dump the barrow on the tow-path. When near the present railroad bridge, when the run was an even grade from the top of the bluff to the tow-path fifteen or twenty feet below, it took good nerves and a steady gait to wheel a load and dump it in the proper place.[18]

Sections Two, Three, Four, Five, Six and Seven of the contracted work all involved construction of the embankment in the river which became the new west bank of the river and the east bank of the canal. The contract for section Two and Three was awarded to a "Mr. Lowell." In a letter to Canvass White dated August 30, 1827, Edwin Douglass reports:

> The work is progressing some faster than it was when you were here. There are about 400 men employed at present. Mr. Lowell has about 120 of them

---

[18] Hayden, *Historical Sketches*, 34.

and is doing tolerable well.[19]

A letter from Douglass to White dated January 15, 1828 reports:

> There is not much doing on the canal at present. The embankment is tolerably secure – as secure as they can be made with shell stone placed on the river side. The weather has been very mild but too wet and muddy to make much progress in making embankments. Mr. Lowell has about 25 hands employed and at present is doing very well. There has been quite an improvement in the building of his bank since you was here.[20]

Unfortunately for Mr. Lowell, his progress was soon interrupted by financial problems. Douglass again wrote to White on March 8, 1828.

> Since I wrote to you last, Mr. Lowell has been under the necessity of stopping his work. He is now trying to make arrangements with his creditors to either take the contract off his hands or to wait on him until he finishes it.[21]

Of the two options described by Douglass, Mr. Lowell chose or was forced into the first. On March 22, 1828, Douglass reported to White:

> There has not been any work done on the River Embankments since last month. Mr. Lowell had been under the necessity of giving up his work and one of his creditors, Mr. Loomis of Suffield, has undertaken it will commence next week.[22]

Alfred Smith also acknowledged the failure of Mr. Lowell in a letter to Canvass White dated March 24, 1828. Smith stated that the Company has relet the embankment of Section Two and Three "but not the slope wall as the latter was declined."[23]

---

[19] Edwin Douglass to Canvass White, August 30, 1827. Canvass White Papers, Rare and Manuscript Collection, Carl A. Kroch Library, Cornell University Library.

[20] Edwin Douglass to Canvass White, January 15, 1828. Canvass White Papers, Rare and Manuscript Collection, Carl A. Kroch Library, Cornell University Library.

[21] Edwin Douglass to Canvass White, March 8, 1828. Canvass White Papers, Rare and Manuscript Collection, Carl A. Kroch Library, Cornell University Library.

[22] Edwin Douglass to Canvass White, March 22, 1828. Canvass White Papers, Rare and Manuscript Collection, Carl A. Kroch Library, Cornell University Library.

[23] Alfred Smith to Canvass White, March 24, 1828. Canvass White Papers, Rare and Manuscript Collec-

Henry L. Loomis was born on June 9, 1794, and died in Suffield, Connecticut on May 6, 1855, at the age of 60. He was the son of Nathaniel Loomis and Bethena Brownson Granger. He married Mary Ballantine King in 1818. Loomis does not appear to have had any more success in completing the contract for Section Two and Three than Mr. Lowell. On September 16, 1828, Canvass White wrote to Henry Loomis:

> The season of the year and the state of the embankment which remains on Section Two and Three, for which you are contractor, requires that the work there should be immediately recommenced. This is not only called for by the necessity of carrying forward those sections so as to complete them seasonably, but especially to prevent the remaining embankment from being carried off in future freshets or floods in the river. The omission, thus far, to renew the prosecution of the work, and the absence of all preparation for so doing, appears to indicate an unreasonable neglect to prosecute your contract, and I hope you will immediately proceed with the same.
>
> Yours respectfully,
> Canvass White, Engineer[24]

The next day, September 17, 1928, White wrote to Alfred Smith that he had recently visited the canal and found no persons employed on Section Two and Three and no appearance of any preparations to renew the work. Based on his observations, he was of the opinion that the work had been unreasonably neglected and he therefore declared Section Two and Three to be abandoned by the contractor Henry Loomis.[25] The declaration of abandonment was a legal step that allowed the Company to proceed with efforts to relet the contract.

The troubles for Henry Loomis did not end there. On December 17, 1828, the Connecticut River Company, acting through its President and its attorney, Alfred Smith, filed suit against Henry Loomis in the Hartford County Court to recover the balances due on three separate promissory notes from Loomis in favor of the Connecticut River Company in the combined amount of $1,594 dollars. The notes were exe-

---

tion, Carl A. Kroch Library, Cornell University Library.

[24] Canvass White to Henry Loomis, September 16, 1828. Canvass White Papers, Rare and Manuscript Collection, Carl A. Kroch Library, Cornell University Library.

[25] Canvass White to Alfred Smith, September 17, 1828. Canvass White Papers, Rare and Manuscript Collection, Carl A. Kroch Library, Cornell University Library.

cuted between Mach 15, 1828 and August 5, 1828. To secure payment of the claim, Smith secured an attachment of three separate properties owned by Loomis in Suffield.[26] The lawsuit raises the question of why the Connecticut River Company was lending money to one of its contractors. One possible explanation is that the first contractor, Mr. Lowell had abandoned the contract in financial distress and possibly owed money to his workers. Perhaps the money was lent so that Henry Loomis could keep the workers on the site so that work could progress. There is no record of how the lawsuit resolved, but the notes were presumably paid, as the attached properties were not levied upon.

Cornish & Company

The same January 15, 1828 letter from Edwin Douglass to Canvass White that mentions that Mr. Lowell is on the site with approximately 25 hands employed also notes that "Cornish & Co. has about the same number" of workmen on site.[27] There is no further mention of Cornish and Company in any of the Canvass White papers.

Philo Bronson, Lyman Atwater, and Elisha Punderson

Two separate court cases were filed on January 2, 1828, charging six Suffield men with trespass for interfering with the contractual rights of two canal contractors.[28] The first contractor was named Philo Bronson. The second contractor was the partnership of Elisha Punderson and Lyman Atwater. The writs of attachment allege that the men, "with force and arms, broke and entered into the plaintiff's chose," on certain lands, and then and there "enticed away the laborers who were at work for the plaintiffs on the canal." Under common law, the term "chose" means rights in property, specifically a combined bundle of rights. In this case, the chose was the right to enter certain Suffield properties to construct the canal, under rights that were conveyed to the Connecticut River Company and assigned to the contractors.

---

26 Connecticut River Company v. Henry Loomis. Kent Memorial Library Collection.

27 Edwin Douglass to Canvass White, January 15, 1828. Canvass White Papers, Rare and Manuscript Collection, Carl A. Kroch Library, Cornell University Library.

28 Philo Bronson v. Daniel Shepard, et. al. January 2, 1828; Elisha Punderson and Lyman Atwater v. Daniel S Shepard, et. al. January 2, 1828, Kent Memorial Library Collection.

There is no specific identification as to which Sections of the canal were assigned to these men, but from the legal description of the property, it can be determined that both properties were in Suffield. Accordingly, Philo Bronson and the Elisha Punderson and Lyman Atwater partnership must have been contracted to build river embankments, as that was the nature of the work let out in Suffield.

Philo Bronson was born in Waterbury, Connecticut on May 15, 1782. His father was Eli Bronson and his mother was Mehitable Atwater. His mother's maiden name suggests that Philo Bronson was a cousin of contractor Lyman Atwater. He represented the Town of Middlebury in the Connecticut State legislature in 1825. He was living with his family in Seneca, Ontario County, New York at the time of his daughter's death in 1834. Since Seneca, New York is also a canal town, with the Cayuga-Seneca Canal opening in 1828 and the nearby Chenango Canal opening in 1833, it is possible he served as a contractor on those canals in addition to the Enfield Falls Canal. Philo Bronson died in Seneca on December 8, 1855.

Lyman Atwater was born in New Haven, Connecticut, on March 3, 1783. He was the child of Revolutionary War Veteran Medad Atwater and Rhoda Dickerman Atwater. He married Clarissa Hotchkiss and they had two children. Atwater was living with his family in New Haven at the time of the 1840 census, in Barkhamsted, Connecticut, at the time of the 1850 census, and in Bennington, Vermont, at the time of the 1860 census. He died in Bennington on March 22, 1862. Atwater was a real estate speculator and frequently involved in litigation, including being the plaintiff in an 1840 collection action where contractor Elisha Punderson was one of the named defendants.

Col. Elisha Punderson was born in New Haven, Connecticut in 1788. He was married to Eunice Gilbert Punderson on February 18, 1817. He died on February 5, 1864 at the age of 75.

Assembling a Labor Force

By using contractors, the Connecticut River Company did not become the direct employer of the canal laborers. Each contractor was responsible for assembling and engaging the labor force needed to carry out his contract. In the canal era, New York City became a clearinghouse for the canal labor force. New York was a natural location for this activity as it was the largest port of arrival for new Irish immigrants. Additionally, when a canal laborer completed a project, unless his contractor retained his services and moved on with his workers to an awaiting next job, the laborer would return to New York in the hope of securing new employment there. Contractors seeking employees would travel to New York to secure the number of men needed.

The Truth Teller was a New York-based periodical published every Saturday by W. E. Andrews and Company from 1825 to 1855. Its target audience was Roman Catholics. Accordingly, it became an effective vehicle to communicate with recent Irish immigrants. This article appeared in *The Truth Teller* on September 22, 1827:

> Emigrants – Groups of persons lately arrived in this country are seen in various directions in our city looking for work. All are anxious to get employment in the city, but there is not one in ten that succeeds; at no period do we remember such crowds of idle people perambulating our streets. We earnestly recommend the new comers to seek employment in the interior of this state and New Jersey. The agents of the Morris Canal, New Jersey, are advertising for workmen. Why do not these emigrants apply immediately and get themselves into good situations before the winter approaches. The Morris Canal Company gives 12 or 13 dollars per month for labourers. By going to Morristown they can get all information and directions to the agents of the canal, at the Summit Level. The cities of Savanah, Wilmington, Charleston, Baltimore, and New Orleans, offer inducements to persons out of employ which should not be overlooked.[29]

The laborers on the Enfield Falls Canal appear to have arrived from New York, which is consistent with the norm for the canal era. One indication in support of that conclusion is that when a laborer was gravely injured on the canal in the summer of 1827, his coworkers travelled to New York to secure the services of a Catholic priest to tend to his spiritual wants.[30] They must have been familiar enough with New York City to know that is where they could find a priest.

A Fee Dispute

Since his appointment as President of the Association for Improving the Navigation of the Connecticut River in 1824, Alfred Smith tenaciously

---

29 *The Truth Teller*, September 22, 1827, W. E. Andrews & Co., New York.

30 Author Unknown, *The History of the Church of St. Mary, Windsor Locks, Connecticut*, 1954.

pushed the Enfield Falls Canal project forward. He carried out his duties with remarkable skill and composure. But the stress of spearheading the ambitions project began to wear on Smith in the late winter of 1828, when he became embroiled in a fee dispute with Canvass White.

It is not unusual for developers and engineers to have disagreements over the course of a project. Design considerations often drive these disagreements. On several occasions during the design phase of the Enfield Falls Canal, Smith had to cope with White's efforts to reverse earlier design decisions. White advocated strongly *(and wisely)* for stone locks after a decision had already been made to use wood locks. When it came to crossing Stony Brook, White originally recommended an aqueduct, then changed to a dam with guard gates, then finally settled back on an aqueduct. White even made an eleventh-hour push to change the design of the entire upper canal (which the Directors almost instantly rejected). Smith patiently and professionally steered the Connecticut River Company through each of these issues. But in February of 1828, following a winter that allowed for little progress on the canal, he received a request for payment from White that tested the limits of his patience.

The offer made by the Connecticut River Company to engage Canvass White included these expectations as to the services he would perform: 1.) White would serve as the Chief Engineer as the Company constructed a canal by Enfield Falls the present year, and 2.) White would procure an examination of the river to "mature and perfect the plans for a general improvement of the navigation between Hartford and Barnet, Vermont." For these services, White would receive the sum of two thousand dollars. The company stated its understanding that White was willing to devote to those services four months of his time "this year" and defray his own expenses. The proposal, although clear in the desire to "complete the canal this year," is silent as to what would happen if the canal was not completed within this time frame.

The acceptance letter by White is clear that the term of the contract is one year. He also agreed to "arrange and digest plans for the improvement of the navigation of the river as may be directed by the Company," but he makes no commitment to examine the upper river as part of that process.

Sometime in the beginning of 1828, Canvass White asked Alfred Smith for a payment toward his services rendered to date of one thousand dollars. In response to this request, Smith inquired of White as to how much time White had actually spent on the Enfield Falls project. White initially eluded answering the question, but finally responded that he had performed between two and three months of service. In response to the payment request, Smith sent only five hundred dollars, not the one thousand requested.

The shortage must have frustrated White. On February 4, 1828, he wrote a letter about his compensation to Alfred Smith. Smith was in Boston at the time lobbying the Massachusetts legislature for approval to advance the Connecticut River Company plans for river improvements in that state. Unfortunately, the letter itself has not survived, but it is referred to in Smith's reply. From the content of Smith's response, it may be assumed that White's February 4, 1828, letter expressed frustration with the reduced payment and requested payment of the balance due of $1500. In response, Smith's suggested that it may be in the Company's best interest to replace White with another Chief Engineer. White was not rattled by the suggestion. Instead, he suggested that the Company hire noted Erie Canal Chief Engineer Benjamin Wright to replace him to supervise the completion of the Enfield Falls Canal. The letter greatly agitated Smith, and on February 23, 1828, he replied to White from Boston:

Dear Sir,

Your favor of the 4th inst. was forwarded to me from Hartford, where it was remailed on the 9th inst. The pressing nature of my engagements here have prevented an earlier reply. I have read your letter with attention. I believe with candor, and will expect a similar consideration of what I now write to you.

Your manner of expressing what you think I proposed to you, is so different from my own expressions, that I am in doubt whether you did or did not take any meaning, as I intended it. If you did understand me correctly, I infer from your letter an expectation on your part, before intimated to, or imagined by me. I am, however, uncertain as to the extent of your expectations. I shall therefore give you my views on several points presented in your letter.

Whether the time spent by you "in walking up and down the river" (as you express it), is all that should be paid for, has never been made a question between us. But supposing the time fairly chargeable by you to our service, should amount to, say two months – Is it your idea that you should receive the full compensation arranged with a view to four months of assistance? In fixing upon a compensation, it was natural to have reference to the amount of service to be rendered. This might be fixed for a particular time, or to superintend the execution of some specific work or works. On our part we calculated upon finishing (or nearly so) the works at Enfield Falls, and also upon some examination by and surveys under you, of certain points where improvements were contemplated above Enfield Falls. This latter was mentioned very early. In the summer, one in particular was specified, and Mr. Dwight & Mr. Hooker came from Springfield (upon my encouraging their expectations) to fix a time for deciding the proper improvement at Willimansett Falls. I mention these things to recall to your recollection what were our expectations and understanding here. The fact that Enfield Falls are so far from having the improvements there completed is certainly no reason why we should expect from you more than the fair and honorable performance of what was mutually understood.

Your remark that you "had flattered yourself that the question as to the number of days spent walking up and down the banks of the river, would not have been agitated." I am not aware that the question has been agitated, in the manner or sense of your remark. Let me ask, - Is no regard to be had to the amount of your time, which is fairly chargeable to our service? Perhaps your remark, just quoted has been drawn forth by the inquiry which I made when you asked me for $1000 dollars. But that inquire was not how much time you had "spent walking up and down the banks of the river." The object of that inquiry was to ascertain how much time you had been with us. I presume my question was so put as to make the impression on your mind that I only thought of the time actually spent here. As you directly replied that you charge us with the time necessary to go and return, and added, from the North river – as well as with the time spent here on our works – I thought my inquiry a proper one, and then as well as afterward when I last parted with you at the steam boat you remarked that you had performed between two and three month of service – in reply to my repeated inquiry.

If it was necessary to give a reason to show the peculiar propriety of my asking that question, it might be found in the method adopted to transact our corporate business. A committee had been appointed in the Spring at my particular request to act on all contracts, both as to making them and disbursing money. Our payments had been made, from time to time, by force of vouchers from that committee to the Treasurer. I anticipated that they might ask me, on my calling for money to be paid you, how much service you had rendered and I put the question I did, to you, that I might be possessed of your view of that matter.

You seem to complain of my asking you to receive 500 dollars when you called for $1000. My object was not to ask delay for the reimbursing of $500 of the 1000 dollars, beyond your next visit. Our treasury was empty. I knew it would accommodate us to pay $500 rather than more. You declined. Our treasurer procured an advance of the 1000 dollars, which was paid to you the same day you asked for it.

A few words as to the unexpected request I made on parting with you the last time. I entered into no discussion whether you were right in saying you should charge us with time not spent here. I did not imagine that you supposed that the time chargeable to us, (between two and three months according to your own view of the matter) laid a foundation for receiving the whole compensation. I did truly suppose that according to your own views between one and two months more of your services would be rendered to us, for the stipulated compensation. I was confirmed in this supposition by my acquaintance with several particulars in regard to your other engagements and because the time mentioned by you exceeded the time actually spent here, and presumed to embrace as much as you yourself thought us fairly to be charged with. I had understood that the Union works remained unfinished and engrossed your thoughts and cares beyond all expectation, when you arranged to aid us, that after your engagement with us you had undertaken the Chief Engineering of the Lehigh Company – that the bridge across the Mohawk, the water power, etc. at the Cohoes, and other causes had necessarily occupied much of your cares. Indeed your letter states that you had fair prospects, when our application was made to you, of being more at leisure than you afterwards found yourself.

It was with the knowledge of facts above alluded to, and under a firm belief that you expected to render us between one and two months more of your services, that I made the suggestion I did on parting with you last fall, by asking you so to distribute the undone of the time for which you were to assist us, as to complete the work as Enfield Falls. I promised that even less than that time would be sufficient for that purpose, if we should relinquish the hope of extending our examination up the river. I was not aware that you would consider this as an unreasonable request. I presume that we shall yet need the assistance, to some extent, of an engineer possessing experience and weight of character. You remarked that much of your time will be required south of the Hudson, the ensuing season, and desire that we would appoint another engineer and recommended Judge Wright.

It seems to me hardly proper, unless from very unforeseen circumstances, that to complete a canal between five and six miles long, like the one at Enfield Falls, a separate engagement should be required with two engineers, of such high standing as Judge Wright and yourself. The reason given by you for delaying your letter also affords a strong reason against our applying to Judge Wright. You say that you delayed to consult him. It is to be presumed that the state of the case, as you presented it to him, is not more favorable to us than as contained in your letter. You seem to imagine ground of complaint against us, the justice of which I can not perceive, and I hope that on further consideration of the subject as connected with the explanations in this letter, that you perceive that your impressions were erroneous, and that your sentiments as explained in the letter of 4 February have done injustice to us. Yet I cannot but suppose that you have communicated those impressions and sentiments to Judge Wright. If that is so, it would form a most serious obstacle to our applying to him, all of us here, have such respect for the character of Judge Wright, not merely as an engineer, but as a man, that we should not readily enter upon such a negotiation with him, while he had an unfavorable, and as we believe erroneous, impression of our treatment of you. As to the skill and qualification of Judge Wright our board entertains the highest opinion.

I repeat that I am in doubt whether you have not misapprehended the object and extent of my request to you made last fall. If you have, I think your unfavorable impression will be reversed. But I wish you to understand most especially, that I never intended to ask of you more for the stipulated compensation that was understood, as well as agreed originally. As to your engagements elsewhere, some of them subsequent to that with us, had engrossed your time and cares much beyond your own expectations. I did believe that you would readily have supplied the deficiency of last year, by such assistance the present year as would bring our works at Enfield Falls to their completion.

When I first received your letter it was hoped that the business in which I appear before the Legislature would have been done by this time. It is not closed. I must therefore write without consulting others and send this with your letter to Hartford for the consideration of our Directors. Two bills are before the Legislature, one for improving Connecticut River, the other for extending the H & H canal toward Vermont. I reserve copies for you of Gov. Clinton's and Col. Baldwin's letter, and other printed Documents.

                      Respectfully
                      Your Obt. Servant Alfred Smith[31]

Smith was worried that the request for $1500 was to satisfy White's initial one year commitment to the project and that a subsequent outlay would be required to secure White's commitment to finish the project. Smith explained that his suggestion to replace White, and to use the funds that would otherwise have been paid to him to compensate the replacement engineer, was only made because he was aware of White's increasing commitment to other projects and resulting reduced availability to devote to the Enfield Falls project. Smith also noted that since

---

[31] Alfred Smith to Canvass White, February 23, 1828. Canvass White Papers, Rare and Manuscript Collection, Carl A. Kroch Library, Cornell University Library.

[32] Canvass White to Alfred Smith, March 18, 1828. Canvass White Papers, Rare and Manuscript Collection, Carl A. Kroch Library, Cornell University Library.

White had discussed the matter with Benjamin Wright, and presumably disparaged the Connecticut River Company in the process, that it would not be in the Company's best interest to negotiate a deal with Wright at that time.

White tried to relieve some of Smith's anxiety when he responded to him on March 18, 1828 as follows:

> …in reply to your first query, I meant the 1500 dollars & expenses for services to be rendered the ensuing season to complete the work at Enfield Falls without regard to former services or payment…I should be left to judge the time that would be necessary to devote to the completion of the work.[32]

Unfortunately, this reply only led to more uncertainty as to what it would take to secure White's commitment to complete the project. Did White mean that his request for $1500 was to secure his future services and the company still owed the $1500 balance of the fee for the first year of service? Or was White acknowledging that the balance of the four months of service he promised would be satisfied in the following year with no additional compensation due?

Despite this uncertainty, Smith took the matter to the Board of Directors for guidance. The Board was equally confused by White's letters, but was also eager to resolve the matter without the inevitable delay and expense that would come with changing engineers mid-project. They instructed Smith to make clear in his reply that the Board was determined to have White finish the project, and then to try to once and for all determine what that would cost the Company. Smith responded on March 24, 1828 with the following:

> Dear Sir,
>
> This evening is the first opportunity which has offered to lay your letter of March 18th before the Board. They have such an opinion of your skill, that they wish to have you see their work finished. Considering the small extent of our works, compared with other, we should have been gratified, if after the explanations in my last, you had thought the original sum of 2000 dollars a satisfactory compensation for your services in seeing the work at Enfield Falls completed provided no more of your time should be required, than we originally stipulated.
>
> You must excuse us – but are not now clear what are your expectations

in regard to the past – whether to claim more than 1,500 dollars & expenses in addition to what you have received, or not. As the first question in my letter was very explicit, we concluded that you would have stated, as distinctly, if you expected any payment besides the sum of 1500 dollars. It is with reference to this view of the subject and of your meaning that the Board authorized me to say they will pay you the sum of fifteen thousand dollars & expenses in addition to the sum formerly paid you, to continue the charge of chief engineer to complete the wors at Enfield Falls, expecting about two months of your time, judiciously decided, as in your letter.

We trust, however, that two months of actual attendance at Enfield Falls, by the chief engineer, will not be needed hereafter, and, shall therefore, propose to make an excursion of a few days, as far up as Miller's Falls for the purpose of a general examination at two or three points on the river. This is anticipated with confidence, so we suffer in a way which cannot be retrieved, for want of your making such examination last season and I add that our not expressing more frequently our wishes on this subject, arose from the short stay, which your other engagements permitted you to make with us; and which prevented you visiting even the nearest point, where improvements are proposed, viz. Willimansett Falls.

The board noticed with pleasure your expressions of unabated interest in the success and prosperity of our operations on the river. In a relation of the kind subsisting between us, it would be obviously unsatisfactory to meet at all unless upon terms of mutual regard and cordiality.

I presume that Mr. Douglass's letter to you will explain whether there is any necessity for an immediate visit from you.

In consequence of Mr. Lowell's failure, we have relet sections two & three – as to the embankment, but not the slope wall as the latter was declined.

Please write to me soon & inform whether we have misunderstood your intentions, and whether the proposal, as herein explained, is satisfactory to you.

     I am with Respect and Regard
     Your Ob. St.
     Alfred Smith, P. C. R. Co.[33]

---

[33] Alfred Smith to Canvass White, March 24, 1828. Canvass White Papers, Rare and Manuscript Collection, Carl A. Kroch Library, Cornell University Library.

CHAPTER EIGHT

The Irish Labor Force

The October 3, 1827 edition of *The Springfield Republican* reported:

> "We are informed that there are nearly four hundred men now at work on the canal to bypass the Enfield Falls, in the Connecticut River."[1]

The workers on the Enfield Falls Canal were predominantly, if not exclusively, Irish immigrants. Windsor Locks historian Jabez Haskell Hayden, who was a sixteen-year-old at the time of canal construction, noted: "The digging of the canal was done by Irishmen who came here for that purpose. I remember having seen but one before."[2]

Hayden's recollection gives an indication of the homogeneity of the local population at the dawn of the nineteenth century. The influx of four hundred Irish immigrants to construct the canal was a demographic curiosity not only to Hayden, but to the entire local population.
In her article, "Traffic on the Connecticut River a Half Century Ago," area resident Nellie Grace Abbe said of the arriving Irish laborers:

> They came on rafts and in scow-boats, often with their entire worldly goods

---

[1] *The Springfield Republican*, Oct. 3, 1827.

[2] Hayden, *Historical Sketches*, p. 33.

knotted up in a red bandana, and many with not even that. They camped in the woods and cooked their food in the huge cauldrons used for melting tar.[3]

The Labor Camps

To accommodate the arriving Irish workforce, The Connecticut River Company first purchased two large tracts of land in what is now Windsor Locks. Here, the Company constructed a crude worker camp. The shantytown was located just south of the land of Charlotte Griswold, on whose land was operated a shad fishery known as the "Griswold Fish Place." The first piece, purchased from Charlotte Griswold, was ten and one-half acres bounded north by the remaining land of Charlotte Griswold, west by what is now Center Street, south by land of Elizabeth, Herlehigh and Harris Haskell, and east by the river. The second piece, purchased from Elizabeth, Herlehigh and Harris Haskell, was seventeen acres bounded north by the first lot above, east by the river, south by the road that led to the ferry, and west by what is now Center Street.[4]

There was likely a second labor camp on the Suffield end of the canal. The most probable location for this camp was on the two parcels of land that the Connecticut River Company acquired from Dan King and Tabitha Simonds in August of 1828.[5] These properties were approximately ½ mile north of the point at which Stony Brook empties into the river. The Simonds piece was the larger of the two, being over two acres in area. The King piece was a narrow strip between the Simonds piece and the river. Simmons acquired her land from King six years earlier. The deed from Simonds to the Connecticut River Company notes that the conveyance includes "the dwelling house and the barn thereon standing being the place formerly occupied by said Tabitha." This language indicates that these buildings were no longer occupied by Simonds at the time of the conveyance. It is possible that the Connecticut River

---

[3] Nelle Grace Abbe, "Traffic on the Connecticut River a Half Century Ago," (*Connecticut Quarterly*, Vol. 3, Hartford, 1897).

[4] Deed, Charlotte Griswold to Connecticut River Company, June 8, 1927, Windsor, CT Land Records, Vol. 31 p. 15; Deed, Elizabeth, Herlehigh, and Harris Haskell to Connecticut River Company, June 8, 1827, Windsor, CT Land Records, Vol. 31, p. 15.

[5] Warranty Deed from Dan King to the Connecticut River Company, August 14, 1828 and recorded September 30, 1828 in Vol. 17, p. 104 of the Suffield Land Records; Warranty Deed for Tabitha Simond to the Connecticut River Company, August 14, 1828 and recorded September 30, 1828 in Vol. 17, p. 105 of the Suffield Land Records.

Company leased the property from Tabitha Simonds and used it to board workers in the existing house and barn, and then subsequently purchased the parcels as it expanded the camp.

Crude wooden boarding houses, frequently called shanties, were constructed within the labor camps as shelters for the arriving laborers. Cooking and cleaning was performed communally by the wives of a small percentage of laborers whose wives joined them in the camp. Members of several longtime Windsor Locks families maintain that upon arrival in Windsor Locks, their immigrant ancestors purchased shanties from the Connecticut River Company and physically moved them from the labor camp. The McKenna family home at the crest of the hill on Center Street across from Pesci Park is believed to have evolved from a labor camp shanty. There is strong evidence that canal worker Lawrence McMahon purchased a former shanty and moved it across the canal and up the bank to the property he had recently purchased from John Morron *(Moron, Moran)*. There he raised his family. A circa 1878 photograph of the canal includes an image of this modest building.

Some of the shanties remained on the site of the Windsor Locks labor camp at least until 1844, when the railroad tracks were laid. A new generation of laborers resided in at least one of the shanties during railroad construction. Railroad subcontractor Thomas Carroll pledged all of his belongings contained in or about the shanty he lived in with his work crew to secure an $875 loan he received from Charles Dillon. The security agreement recorded in the Windsor Land Records gives a detailed description of the contents of the shanty, and in so doing provides a glimpse into life in the labor camp. The document states that to secure his debt, Carroll does,

> ... hereby bargain, sell, assign, transfer, and set over to said Dillon the following articles of property situated in said Windsor and used in and about the performance of said contract and now in my possession and occupancy, viz: 5 horses, 5 harnesses for same, 5 carts, 15 pick axes, 30 shovels, one barn situated on the land of Widow Griswold in said Windsor now occupied by me, one house or shanty on said land occupied by me, 13 beds and bedding in said shanty and 2 bedstands, 1 bureau, 3 stoves, six chairs in said shanty and used in housekeeping, 11 barrels of beef, 1 barrel of pork, 10 barrels flour used for the family of the subscribers and used in or near said shanty, and all other of the same belong to said Thomas, about 4 tons of hay, about 40 bushels of oats and 20 loads of manure situated in or near said barn, 2 kegs of powder, 4 crow bars, 4 chisels and 1 saw, situated in said

Windsor aforesaid and used on or about said work.⁶

Disease was the biggest risk facing canal laborers living in the labor camps. The following article appeared on September 28, 1826, in *The Pittsfield Sun* about an incident on the Delaware and Raritan Canal. That canal bisected New Jersey primarily to bring coal from the Pennsylvania anthracite coal fields to the New York and New England markets. It was designed by Canvass White, the same engineer who served as the Chief Engineer on the Enfield Falls Canal.

> A few days ago, four Irish laborers upon the Delaware and Hudson Canal, were found dead in one of their "shantees" upon the line of excavation, a few miles above Trenton, N.J. The immediate cause of their decease is not known to a certainty, but they are supposed to have been simultaneously attacked with the Cholera, and were only enabled to reach their beds, where they were subsequently found in a state of putrefaction. It was not until after a considerable time that any person could be found to perform the last office of interment. At length some colored persons were induced, by the offer of a large reward, to undertake their burial. These having dug a pit to windward of the cabin, succeeded in interring them, and then setting fire to the "shantee," they burnt it to the ground.⁷

The New Orleans New Basin Canal commenced construction in 1833. Disease killed workers with such frequency on that project that builders expressed a preference for Irish labor over slave labor. The reasoning behind that preference was that a dead Irishman could be replaced in minutes at no cost, while a dead slave resulted in the loss of a several hundred-dollar investment.⁸ This rhyme of unknown origin is based on that calamitous canal project.

> Ten thousand Micks, they swung their picks
> To dig the New Canal
> But the Choleray was stronger than they

---

6 Warranty Deed, Dan King to the Connecticut River Company, August 14, 1828 and recorded September 30, 1828 in Vol. 17, p. 104 of the Suffield Land Records; Warranty Deed, Tabitha Simond to the Connecticut River Company, August 14, 1828 and recorded September 30, 1828 in Vol. 17, p. 105 of the Suffield Land Records.

7 *Pittsfield Sun*, September 28, 1826.

8 Mary Helen Lagasse, A Call to Remember 8,000 Irish Who Died While Building the New Orleans Canal, Irish Central, Nov 07, 2013.

And twice it killed them all.

Occupational Hazards

Aside from the dangers of life in the labor camp, working conditions on the Enfield Falls Canal were harsh and dangerous, just as they were on all canal projects during America's canal building era. Author Peter Way explains in his excellent book *Common Labor, Workers and the Digging of North American Canals 1780-1860*:

> Rock excavation was the most dangerous task, as blasting was the only way to cut through solid stone. The procedure seemed deceptively simple: a hole was hand-drilled and packed with powder, and a fuse was inserted at the top. Yet it took three men using a rock chisel and a sledgehammer a day to drill a one-and-three-quarter-inch hole twelve feet deep, or a three-inch hole four feet deep. Fuses were a matter of guesswork, strips of match-paper soaked in a saltpeter solution that when dried would burn according to the strength of the solution. A blaster would light the fuse and run for cover. Some charges failed to go off or merely rumbled the ground, but if all went well, rock fragments would fly into the sky and then rain down, and when the dust had settled the rock bed would have receded several feet. Overcharging could propel boulder-sized rocks hundreds of yards through the air to land on cowering laborers and crash into shanties or nearby communities.[9]

Social Isolation

There is little evidence of interaction between the laborers and the local population, and their existence was that of a transient workforce. There was no effort from the local population to mix with the Irish laborers, and there was no effort made by the workers to blend into the local population. The expectation of both was that the Irish would move on to the next project as soon as the present project was complete. "Relatively few of the nameless Irish workers, who lived near their work in temporary board houses or shanties, stayed after the canal opened in November 1829."[10]

The Canal's Impact on Early Catholicism in North Central Connecticut

In addition to being Irish, the majority of the Enfield Falls Canal work-

---

[9] Peter Way, *Common Labor* (Cambridge, Cambridge Uiversity Press, 1993).

[10] Raber and Malone, Connecticut Canal Historic Exhibition Center Feasibility Study and Master Plan, June 1991.

force was Catholic. Although Hayden's comment that he had seen but one Irishmen before refers to the country of origin, odds are that he had seen even fewer Catholics than Irishmen. Tolerance for papists *(Catholics)* was not widespread at this time in the nation's history and the attitude of the local population frequently moved beyond passive intolerance to active prejudice. *The Litchfield Post* reprinted the following article from *The Christian Secretary* on November 1, 1827:

> Roman Catholic Superstition in Connecticut – There are about 400 Irishmen at work on the canal at Enfield Falls. Most of them are Catholics. A few days since one of them died. He had no priest by him to receive his confession, and give him absolution before his death. Immediately after his decease, three or four of his countrymen engaged an inhabitant of Suffield to convey the body to Albany, New-York where there was a Catholic Priest. These friends accompanied the corpse in the same wagon. On their arrival in Albany, for 30 dollars the priest gave absolution to the soul and body of the dead man, in the name of the Father, Son and Holy Ghost, the Virgin Mary, the Holy Angels, and the Spirits of the departed Saints. So ended the scene of spiritual wickedness in high places. And the surviving parties returned, satisfied that the soul of their friend was safe and happy.[11]

This is the cultural milieu into which the Enfield Falls Canal laborers entered.

The earliest indication of any devout Catholic population in North Central Connecticut can be traced to February 1823 when Rev. John Cheverus, the first Roman Catholic Bishop of the Diocese of Boston, whose jurisdiction included all of New England, wrote the following inspirational letter:

Boston, Feb. 7, 1823

To the Roman Catholics residing at and near Hartford.

My beloved friends and children in Jesus Christ,

> Your letter of the 3d instant has been duly received, and has afforded me great gratification. I wish I could go immediately and pay you a visit, but it is out of my power to go till after Easter. I shall give you notice a

---

[11] *Christian Secretary*, reprinted in the *Litchfield County Post*, November 1, 1827.

fortnight before my going. In the meantime, you will do well to procure a room and meet every Sunday to perform together your devotions. Let one who reads well, and has a clear voice, read the prayers at Mass, a sermon, or some instruction out of a Catholic book. If you are destitute of books, let me know, and I shall send some by the first opportunity. During the ensuing Lent which is to begin next Wednesday, flesh-meat is allowed on Sundays, Mondays, Tuesdays and Thursdays, except the last or holy week, but only once a day except on Sundays.

I am happy to hear that you openly profess your religion. Never be ashamed of it, nor of its practices, and above all do honor to it by an irreproachable conduct. Be sober, honest and industrious; serve faithfully those who employ you, and show that a good Catholic is a good member of society, that he feels grateful to those who are kind to strangers, and sincerely loves his brethren of all persuasions, though he strictly adheres to the doctrines of his own Church.

It is thus, my beloved friends, that you will silence prejudice and win the esteem and favor of all the inhabitants of this hospitable country.

Be assured that nothing I can do will be wanting on my part to promote your spiritual welfare. At my first visit we may fix upon regular periods when one of my Reverend brethren, or myself, will go to administer to you the sacred rites of our holy religion.

With affectionate feelings of paternal regard, and fervently imploring upon you all the blessings of the Father, the Son, and the Holy Ghost.

I remain your friend and pastor.

John Cheverus, R. C. Bishop of Boston[12]

Sadly, Bishop Cheverus never made it to Hartford. Unknown to him at the time he wrote to the Hartford Catholics, he had already been recalled to France by Pope Pius VII after twenty-seven years in Boston. Bishop Cheverus' successor, Rev. Benedict Joseph Fenwick, kept his predecessor's promise, but distance, inferior modes of transportation, and an abject shortage of priests, prevented him from doing so for several years after Cheverus first made his promise.

Father John Power of New York City was the first known Catholic priest to minister to the Irish canal workers in Connecticut.

---

12 *United States Catholic Press*, Boston, February 7, 1823.

In August 1827, a laborer was seriously injured working on the canal and his fellow workers sought out a priest to administer the last rites of the Church to the dying man. Because these Irish-Catholics had been brought from New York and were probably quite unaware they were within the jurisdiction of the Archdiocese of Boston, a man was sent to New York and returned with Reverend John Power, native of County Cork, Ireland, and Vicar-General of the Diocese of New York. Father Power was also the Pastor of St. Peter's Church in Lower Manhattan, which has been frequently depicted recently in the media due to its location across the street from the World Trade Center site. The trip was accomplished by boat, as regular steamship service between Hartford and New York was then available. As reported in the history of St. Peter's Parish, after performing his work of mercy, Reverend Power said mass in an open field.[13] Later historical accounts identify the site as an open field beside the river, under a canopy of trees, and near the head of the shad fishery.[14] Such descriptions place the site of the first Roman Catholic mass in Windsor Locks in the vicinity of the workers camp, most likely under the stand of trees along the southern border of the Griswold land. On his return to New York, Reverend Power said mass and baptized some children in Hartford and visited with Catholics in Wapping *(now South Windsor).*[15]

At the urging of the canal workers, Father Power returned to the labor camp in October 1827, where he preached, said mass and conducted a sort of mission. On one of these visits, he celebrated Mass in New Haven on his return trip, having missed his connecting steamer there.[16]

The Catholics on the canal went a year without spiritual guidance until Reverend R. D. Woodley was ordained in Boston and immediately assigned the entire States of Rhode Island and Connecticut as his territory. Reverend Woodley made at least three visits to Hartford and to the Canal during September 1828, May 1829, and July 1829. *The Catholic Press*, published between 1829 and 1832 in Hartford by

---

[13] Thomas Duggan, *The Catholic Church in Connecticut*, (New York, The States History Company, 1930),

[14] Unknown, *The History of the Church of St. Mary* Windsor Locks, Connecticut, 1954; Unknown, *The Story Windsor Locks*, 1954, p.41.

[15] Leo Raymond Ryan, *Old St. Peter's, the Mother Church of Catholic New York (1785-1935)*, (New York, United States Catholic Historical Society, 1935), p. 168.

[16] Ibid

A. M. Talley, printed this article regarding Rev. Woodley's visit on July 12, 1829:

> The Rev. Mr. Woodley returned to this city on Sunday afternoon the 12th inst. from his visit to the Catholics on the Enfield canal. Notice of his arrival amongst them in that quarter, having been circulated, a wide concourse of all denominations attended public service in the forenoon of Sunday. He baptized several children, admitted a considerable number to the Holy Communion, and a greater number received the sacrament of penance. On his return to this city he baptized several children. He set out Monday for New Haven, New London, and Providence.[17]

The baptismal records from Rev. Woodley's three visits to the area survive in the archives of the Archdiocese of Boston in a leather-bound ledger written in his own hand. In addition to providing the names of many canal workers, the records are significant in that they prove that several of the workers had wives and families with them. Historical accounts of other canal projects of the era indicate that wives were often on the payroll of the canal contractors, employed as cooks and laundresses. Although the exact location of Rev. Woodley's visits is not stated, the most likely site is the canal worker camp as the visits are reported to have taken place in what is now Windsor Locks.

The following baptisms are recorded in Rev. R. D. Woodley's register:

On the canals about Hartford. September 1828

Baptized Mary, about a week old. Parents Jno. & Mary Sliney. Sponsors Jas. Hennessey & Catherine Shannon

Baptized Ann, about 1 mo. old. Parents Austin & Ellen Doyle. Sponsors Michl. Furley & Mary Travers

I supplied ceremony of baptism to Thomas, about 4 wks. Old. Parents Owen & Mary Sullivan. Sponsors Jno. Barry & Ellen Lynch

I supplied ceremonies to Margaret 1 week. Parents Thos. & Marg. Haly. Sponsors Jas. Hayes and Marg. Manning

I baptized Mary born Aug. 10, 1828 Par. Wm & Bridget Doran. Sponsors Jms. Pelfory & Mary O'Neill

---

17 *The Catholic Press*, July 18, 1829.

Eleanor, Born 25th inst. Parents Pat. & Ellen Walsh. Sponsors Jno. Glughan & Elizabeth Cardle

John, born 7th April 1828. Par. Danl. & Jerusha Grinnel. Sponsors Jno. Couron & _____.

Patrick about 4 weeks. Parents Hugh & Mary Curry. Sponsors Wm. Cosler and Bridgit Smith.

Enfield or Hartford Canal (1829)

May 20th
Baptized John 3 months old. Par. Garret & Johanna O'Brien. Spn. Rich. Butler and Cath. Lannigan

Baptized Margaret 4 months. Par Jno. & Mary Hayes. Spns. John Holden & Ellen Brennan

May 21st
Baptized Dennis 4 months. Par. Tho. & Mary Fitzpatrick. Spns. Rich. Jenks and Mary O'Connel.

Baptized John, 6 weeks. Parents Jno. & Mary Murray. Spn. Daniel Murphy & Mary Sliney.

Connecticut on the Enfield Canal (1829)

July 10
Baptized John. 4 weeks old par. Michl & Mary FitzGerald. Spons. Levi O'Dee & Bridgit Walsh

Baptized Mary 2 months old (illegit). Par. Ellen Fox spon Thom Howard & Mary Hayes

11th
Baptized Ellen 4 day old pars. Michael & Ellen Farrel. Spns. Jno. Cregan & Mary O'Neill

Bishop Fenwick himself came to Hartford in July 1829 and said mass at the offices of *The Catholic Press* on Main Street in Hartford before preaching to a multi-denominational audience at the State House. While there he negotiated the purchase of an abandoned Episcopal Church and a lot on Talcott Street to move it to. Reverend Woodley completed the sale in 1829, when most of the workforce was engaged in Suffield.

On August 26, 1829, Reverend Bernard O'Cavanaugh became the first Catholic priest to take up residence in Hartford and was appointed pastor of the small congregation in Hartford and missionary priest for the entire state of Connecticut. Almost immediately, he set out to meet his flock. On September 5, 1829, the following notice appeared in *The Catholic Press:*

> The Reverend Bernard O'Cavanaugh left this city on the 2nd inst. for the purpose of visiting the Catholics on the Enfield Canal, during his stay amongst them, he baptized several children, heard a considerable number of confessions, admitted some to the Holy Communion, and received some liberal donations toward the liquidation of the debt contracted by the Catholics of Hartford for the Church and Lot, lately purchased by them in this City. We are authorized to state, that special thanks are due to Col. Norris, for the very polite and marked attention which he exhibited towards our Pastor on the occasion of his late visit, both, in furnishing him with an excellent apartment in his hospitable mansion, and permitting him to preach and celebrate therein, the devine mysteries of the most holy and adorable sacrifice of the Mass. In his discourse, the preacher adverted to the foul and calumnious charge entertained by some of his fellow citizens that Priests forgive sins for money. He refuted the slanderous charge by admitting the objection that no one, but God can forgive sins; yet proving that Almighty God has and does make use of the instrumentality of men, and is pleased to forgive sins committed after baptism, by means of contrition, confession, satisfaction, and the Priest's absolution; and that all this is effected without receiving one single, solitary, American cent. The Rev. gentleman returned on last Thursday afternoon.[19]

Col. Samuel Norris was Protestant. He purchased an eighty-five-acre farm in Suffield in 1806 from David Kendall, bounded easterly by the river about where the dam would eventually be built. The land crossed what is now East Street and his "hospitable mansion" appears to have been located on the West Side of East Street, just South of the intersection with Canal Street. He served briefly in the War of 1812 as part of the Connecticut Militia from September 13, 1813 to November 1, 1813 attaining the rank of First Lieutenant. He rose to the rank of Colonel following the war. He was elected to several municipal positions, including the position of Assessor in 1822, and he also served in the state legislature. Following his death on July 1, 1853, he was buried in the Old Center Cemetery in Suffield. His wife Nancy joined him there following her death on November 13, 1863.

One can only speculate as to whether the generous accommo-

---

[19] *The Catholic Press,* September 5, 1829.

dation afforded to these early Catholics by Col. Norris and his wife was the result of some spiritual curiosity or merely the result of genuine and respectable hospitality. There is perhaps some irony to be found in the contrast between the spirited defense to Protestant criticism of Catholic rituals contained in Father O'Cavanaugh's sermon and the charity and respect demonstrated by the good Protestant Colonel himself in offering up his own parlor to serve as pulpit. Surely, his kindness was appreciated by all canal workers who attended the Catholic services he hosted. Norris served as lock-tender at the guard lock adjacent to his property during the first year of canal operations.

Canal Worker Burials

The crude Enfield Falls Canal laborer camp in the part of Windsor that is now Windsor Locks is the site of the earliest celebrations of the Catholic Mass in north central Connecticut. It is also the site of some of Connecticut's first Catholic burials. Local historian Jabez Haskell Hayden reported in 1900 that most of the immigrant laborers left the area after construction.[20] Hayden's observation appears likely, as the canal construction labor force was largely itinerant. A small number of laborers remained after the November 1829 opening of the canal, a fact that jibes with the historical precedent on other canal projects that a small number of laborers were often retained after construction to perform ongoing maintenance.

In addition to the dying worker whose spiritual needs were attended to by Reverend John Power in August 1827, who is presumed to have been buried near the labor camp upon his ultimate demise, the gravesites of other canal workers, some bearing headstones, are in Suffield's Old Center Cemetery. Among them is Michael Costello, of Kilfinanne, County Limerick, Ireland *(who we were introduced to earlier in this work)*. Costello's headstone indicates that he died on June 24, 1829, at the age of 36. His last name is spelled Costlo on the headstone. The headstone also indicates that it was erected by a man named James Hannassay, which is a name that matches one of sponsors to a baptism performed by Reverend R. D. Woodley at the labor camp in September 1828.

---

[20] Hayden, *Historical Sketches*, 34

Other canal worker burials in the Old Center Cemetery are Timothy McMahon, who died on December 11, 1827 at the age of 27. His gravestone notes that it was erected by his brother Cornelius McMahon. A search of immigrant ship passenger lists reveals that 26-year-old Timothy McMahon arrived in New York City on July 16, 1827 aboard the immigrant Ship *Cortes*. His arrival coincides with the period contractors who had secured work on the Enfield Falls Canal would have been in New York seeking laborers to fulfill their contracts.

Next to the grave of Timothy McMahon is the grave of John McMahon, of Limerick, County of Limerick, Ireland. John McMahon died on December 12, 1835 at the age of 24. The headstone indicates that it was erected by his brother Lawrence McMahon. The death of John McMahon occurred more than five years after the canal was considered to have been completed. John McMahon and his brother Lawrence McMahon frequently appeared in the records of the Connecticut River Company from 1830 to 1833. These Irish workers were retained following completion of the canal to perform troubleshooting, repair and maintenance tasks.

Although it is clear from the evidence that Cornelius and Timothy McMahon were brothers, and that John McMahon was the brother of Lawrence McMahon, it is not known with certainty if all four men were brothers. Since John McMahon's gravestone indicates that he was 24 years of age at the time of his death in December of 1835, then it is likely that he was born in 1811. A baptism record from the Catholic Parish of Oola and Solohead (which straddles the County Limerick and County Tipperary border) indicates that a John McMahon was baptized on July 13, 1811. His parents were Cornelius McMahon and Ann Ryan. John is a common given name, but Cornelius is a much less common name. The combination of a John McMahon from Limerick, matching the age of the John McMahon buried in Suffield, with a father named Cornelius McMahon, greatly increases the odds that John McMahon of Oola is the same man buried in Suffield. It also increases the odds that Timothy and Cornelius McMahon are of the same family as John and Lawrence McMahon.

Several additional unmarked gravesites are in the same section of the Old Center Cemetery as Costello and the McMahons. Although these Catholic burial sites are in the center of the cemetery today, the cemetery has expanded since these burials. At the time of the burials, the gravesites were relegated to the rear row of the cemetery.

As noted previously, baptism records in the archives of the Archdiocese of Boston indicate that Reverend R. D. Woodley came to Windsor Locks on three occasions in 1828 and 1829 and baptized fifteen infant children of canal workers. The infant mortality rate for the general population in Connecticut was about fifty percent in the 1820s. The death rate for the children of immigrant laborers living in rudimentary shanties must have been even higher. It is almost a certainty that some of these canal era babies did not survive their time in Windsor Locks.

Statistical compilations from other canal construction projects of the era paint a grim picture of canal life. Worker deaths from disease and occupational hazards were commonplace. In one case, during construction of the Wabash & Erie Canal through Huntington County, Indiana, it is estimated that one worker died for every six feet of canal completed. Although such reports may be exaggerations, even if workers died at one-one hundredth of that rate, the five and one-half mile Windsor Locks Canal would have experienced nearly fifty worker deaths. Aside from those few burial sites in Suffield, where are the remains of these poor souls buried?

A review of a curious restrictive covenant in a deed from the Connecticut River Company to the Hartford & New Haven Railroad Company fifteen years later sheds some light on this mystery. After the canal was completed in 1829, its operators enjoyed fifteen years of competition free prosperity. The emergence of railroads, with their speed and efficiency, eclipsed the canal's utility. In 1844, the operator of the canal, The Connecticut River Company, received $3,500.00 to convey the land for the railroad along the side of the canal to the Hartford & New Haven Railroad Company. The deed of conveyance includes several parcels of land along the west side of the canal. One of those parcels was a small seventy-foot by seventy-foot piece on the north side of Carlton's basin. The piece is described as stretching thirty-five feet to each side measuring from the centerline of the railroad bed. The labor camp was located on the south side of Carlton's Basin. Carleton's basin, also referred to as the lower basin, was a body of water fed by the brook running behind what is now Pesci Park. Originally, the basin opened to the canal and boats docked in the basin to load lumber and unload other goods. When the railroad was laid out along the west side of the canal, it crossed the neck of the basin on pilings.

In the early 1900s, the basin was filled in. A culvert was constructed to carry water from the brook to the canal. Windsor Locks Commons now sits where the basin was located. The legal description of the small piece to the north of the basin contains a curious feature. The conveyance contains a restrictive covenant that ensures:

> **No excavation shall be made on said strip of land, deeper than is necessary for properly grading and draining said road or way.**[21]

No such restriction was included with regard to any of the other parcels conveyed.

It can reasonably be concluded that this small parcel on the west bank of the canal and the north bank of the basin is the burial site of dozens of canal laborers and their family members. This conclusion is supported by the fact that the remains of workers have been discovered along the banks of several canals constructed during the same era. Additionally, both the labor camp and the burial site are on parcels that were owned by the Connecticut River Company, in close proximity to each other.

It should be noted that various local historians, including the late Hugh Starr, have advocated that a Catholic cemetery existed in Windsor Locks in the area between Windsor Locks Commons and the railroad tracks. This site is to the immediate west of the Irish canal worker burial site. The discovery of various headstones in the area as well as an identification of the cemetery in a 1935 inventory of State cemeteries compiled by the Works Progress Administration gave rise to this contention.

The existence of a Catholic Cemetery in this area has been disproven. Although the site is wooded now, that was not always the case. Dr. Sidney Burnap maintained a right of way over this site that led from his mansion abutting the site to the north and ran beside the railroad tracks to the station and freight house. Several photographs of the well-groomed lawn and cinder walkway on the site have recently come to light. There are no gravestones, standing or otherwise on the site. Additionally, one of the headstones located on the site in the 1970s bears the name of Eliza Nugent whose name also appears on a subsequently placed Nugent family stone in St. Mary's Catholic Cemetery on Spring Street.

---

21 Deed, Connecticut River Company to Hartford Railroad and New Haven Railroad Company, 1944.

For many years, a man named Dan Leach operated a coal yard on the Windsor Locks Commons site. Census records show that the father of Dan Leach was a stonecutter and headstone engraver. It is possible that the stones found on the site believed to be an old Catholic cemetery were merely discarded or superseded stones disposed of behind his son's coal yard by stonecutter Leach.

One can understand how the existence of a consecrated burial ground *(now under the railroad bed)* alongside the canal for Irish Catholic canal laborers who died during construction and the discovery of stones on an adjacent site might have confused the issue *(even though they bore dates of much later deaths)*, and led some to the conclusion that the two sites were one in the same. For the reasons stated above, that is not the case.

Other Means of Identifying Canal Laborers

Although the largest share of laborer names that have been identified come to us through baptismal records, burial records, and Connecticut River Company disbursement registers, occasionally workers are identified through more random sources.

Writs of Attachment

In the previous chapter, two writs of attachment, charging six Suffield men with trespass for interfering with the contractual rights of two canal contractors, were reviewed. They included the names of three contractors. One of the writs also identifies the name of one of the Irish laborers who was allegedly enticed away from his work on the canal.

The writ filed by Elisha Punderson and Lyman Atwater recites that the defendants, "then and there disturbed plaintiff in his work upon the Canal by enticing James Hayes and others, his workmen, to leave the work upon the Canal and by disturbing and threatening plaintiff's workmen and other wrongs damaging to the plaintiffs."[22] The name of laborer James Hayes also appears in Rev. Woodley's baptismal register.

Newspaper Account of Stolen Money – Thomas and Jeremiah Casey

---

[22] Philo Bronson v. Daniel Shepard, et. al. January 2, 1828; Elisha Punderson and Lyman Atwater v. Daniel S Shepard, et. al. January 2, 1828, Kent Memorial Library Collection.

The July 15, 1828 *Hartford Courant* contains the following article:

MONEY STOLEN

On Sunday last, at the works of the Connecticut River Company near Enfield Falls, the trunk of the subscriber was broken open, and the following money stolen, viz: - 17 Sovereigns, 23 dollars in bills on the Hudson and Delaware Canal Company, one 5 dollar bill of the Phoenix Bank, Hartford, and about 7 dollars in Spanish mill dollars. The above money was supposed to have been taken by Thomas Casay, a man about 25 years of age, about 5 feet 8 inches high, and thick set. Whoever will give the subscriber such information as will enable him to recover the money, shall be suitably rewarded.

JEREMIAH CASAY
Windsor, Con. Warehouse Point, July 1. [23]

This article identifies at least one and possibly two canal laborers. Additionally, it underscores the difficulty of carving out one's existence in the canal laborer camp. Lastly, it should be noted that some of the money stolen was in the form of bills of the Hudson and Delaware Canal Company. These notes could only have been earned by working on the Delaware and Hudson Canal, which was completed just as construction on the Enfield Falls Canal was commencing. This information is consistent with our understanding that the canal laborers on the Enfield Falls Canal were part of an itinerant work force.

Did Any Canal Workers Remain in the Area Following Completion of Canal?

Jabez Haskell Hayden's *Historical Sketches* is a collection of reminiscences, or "fugitive sketches," as Hayden called them, written in the last decade of his life. Most of the sketches had been published in *The Windsor Locks Journal* in the decade prior to Hayden's 1902 death at the age of 91. In 1900, the Journal Publishing Company arranged and published *Historical Sketches* as a collected work.[24]

In one of his sketches, first published in *The Windsor Locks Journal* on January 13, 1899, Jabez Haskell Hayden speculated as follows

---

23 *The Hartford Courant,* July 15, 1828.

24 Hayden, *Historical Sketches,* p. 34.

about the Irish canal laborers:

> Whether any of the present residents of Windsor Locks are descendants of those first comers I am not certain, but I think the increase of the Pine Meadow population has all come in with the increase of mills and manufactories, the result of opening the canal which was originally created for the purposes of navigation.[25]

Perhaps Hayden was unaware that just a year before he wrote those words, Windsor Locks physician James Coogan, whose father, James, settled in Windsor Locks in 1845, had already answered the first question of whether any canal workers settled in the area. Coogan wrote a chapter entitled, "The Irish Element in the Windsor Towns," which appeared in Henry Reed Stiles's 1898 work, *The History and Genealogies of Ancient Windsor, Connecticut*. Coogan wrote:

> The digging of the canal at Windsor Locks attracted a number of Irishmen – and the names, Guinney, Moore, Burke, McMahon, Fitzgerald, Hayes and Doyle testify to their race and the land of their birth. Several of these became prominent residents – thus forming the nuclei of those Celtic homes which now total one-third of the total within the boundaries of the old town of Windsor.[26]

Coogan's recollection is supported by the 1830 Federal Census and by the Connecticut River Company Records from 1830 to 1833 maintained by the Connecticut Historical Society.[27]

Among the residents of Windsor *(Windsor Locks was not separately incorporated until 1854)* recorded in the 1830 Federal Census are Mitchel *(Michael)* FitzGerald, Clarence *(Lawrence)* McMahon and Edmund Burke and their families. The FitzGerald household contained one male and one female between the ages of thirty and forty and one male child under the age of five. The McMahon household contained one male and one female between the ages of twenty and thirty and a female

---

[25] Ibid.

[26] James Coogan, "The Irish Element in the Windsor Towns," in, Henry R. Stiles, *The History and Genealogies of Ancient Windsor, Connecticut* (New York, C. B. Norton, 1859; republished Somersworth, NH, The New Hampshire Publishing Company, 1976), p. 775.

[27] Connecticut River Company Records, Connecticut Historical Society, Edgar F. Waterman Research Center.

child under the age of five. The Burke household contained a male and female between the ages of thirty and forty and two male children under the age of five. Among the residents of Suffield recorded in the 1830 Federal Census are Michael Hayes and Martin Doyle. Hayes' household consisted of one male and one female between the ages of thirty and forty and one female under the age of five. Martin Doyle's household included one male between the ages of thirty and forty, one female between the ages of forty and fifty, one female between the ages of twenty and thirty, and two females under the age of five. In both the census for Windsor and Suffield, by reviewing the names of the families living in close proximity to these young Irish families, it can be ascertained that each of these families lived near the canal.

Coogan's listing of the Irish families who came to work on the canal and remained in the area following the canal era is further supported by the financial records of the Connecticut River Company for the first four years of operation.[28] Among these records are receipts for payments made between 1830 and 1833 to the following workmen: Michael Fitzgerald, Martin Doyle, John McMahon, Lawrence McMahon, Michael Guinane, Edmund Burke, Michael Moore, John Hayes, Joseph Hayes and Daniel Hayes. These names all match names listed in Coogan's article. Additional names appearing in the Connecticut River Company records from 1830 to 1833 are John Cragan, Thomas O'Connell, Daniel O'Connell, John O'Neill, Michael Killbride, Philip Killbride, John Lane, Michael Farrelly, John Murray, Frank Gillday, Patrick Timmons, Daniel Deland, Michael Halley, Timothy Cain (Kane), Michael Connery, Hugh Murphy, Felix Maguire, James Kennan, William Towers, John Donoho, John Maloney, Michael Henderson, William Brunson, Thomas Collins, James Bostwick, and Edward Ryan. It may be presumed that this second grouping of Irish canal workers did not remain in the area following construction since they are not listed in Coogan's article, nor do they appear in subsequent area records.

Michael Guinane (Guinney, Guinang)

Although neither the name Michael Guinane nor any variants of it appear in the 1830 Suffield or Windsor Census, the name appears in both

---
[28] Ibid.

Coogan's article and the Connecticut River Company pay records from 1830 to 1833. A Michael Guinane does appear in the 1850 Windsor census. In that census, he is reported to be 47 years of age and living with his wife Caroline age 42 and five children, Eliza, John, Thomas, James and Mary, who range in age from six to seventeen. The fact that Michael Guinane does not appear in the 1830 or 1840 census does not mean he did not live in the area. It was not until the 1850 census that household members were listed. If he was living with another family or landlord, his name would not appear. His residence in 1850 is near John Morron, who owned the land surrounding the upper canal basin, which puts him in proximity to the canal. His age matches the profile of a canal worker. Further, Guinane is not a particularly common Irish name, so the odds of one Michael Guinane being in Windsor in 1830 working on the canal and a different Michael Guinane living in the area in 1850 is unlikely.

In the 1860 Windsor Locks census, a Michael Guinang, age 59, is living in the same area with his wife Caroline and children Thomas, James and Mary. In the 1870 census Michael Guinang, age 70, is listed as living with Caroline Guinang age 61. His occupation is reported as "works on canal." In the 1880 census, Caroline Guinang is reported to be living alone. The absence of Michael leads to a presumption that Michael died between 1870 and 1880, and, in fact, there is a gravestone at St. Mary Cemetery in Windsor Locks bearing the name M. Guinang. The inscription notes that M. Guinang was born September 29, 1800, and that he died September 25. 1870.

John McMahon

The name *(using several variants of McMahon)* appears in many of the Connecticut River Company pay records from 1830 to 1833. During 1830, John McMahon was paid by contractor Abiel King, but in the following years directly by the Connecticut River Company. He does not appear in the 1830 U. S. Census, most likely because he was boarding with another person at the time. Only heads of household were reported by name in the 1830 census. John McMahon died on December 1, 1835, and is buried in the Old Center Cemetery in Suffield, Connecticut. His gravestone notes that he was just twenty-four years old at the time of his death.

Lawrence McMahon

This name *(using several variants of both Lawrence and McMahon)* appears in both the 1830 and 1840 U. S. Census. In the 1830 census the household contains one male and one female between age twenty and thirty, and one female under the age of five. That census also reveals that the McMahon household, along with those of canal workers Edmund Burke and Michael FitzGerald, was just south of the home of Mrs. Griswold *(whose family operated the shad fishery)* and just north of the home of Samuel Denslow. This suggests that McMahon, Burke, and FitzGerald were living in three separate shanties in the laborer camp in 1830. In the 1840 census, Lawrence McMahon's household contains one male and one female between age forty and fifty, one female aged ten to fifteen, one male aged five to ten, and two females under age five.

The Hale Index of Connecticut headstone inscriptions indicates that both Lawrence McMahon and his wife Johanna McMahon are buried in Holy Trinity Cemetery in Hartford.[29] Lawrence's headstone indicates that he died on April 4, 1844, at the age of forty-four. Johanna's headstone indicates that she was born on May 12, 1803, and died on January 22, 1860. It makes sense that they were buried in Holy Trinity Cemetery because at the time of Lawrence's death in 1844, it was the only consecrated Catholic cemetery in the area, other than the canal worker burial ground next to the laborer camp, and that burial site had been closed and railroad tracks were being laid directly over the top of it that very year.

Parish records from St. Munchin's Parish in Limerick City, Ireland indicate that Lawrence (Laurence) McMahon married Johanna McCarthy on May 21, 1828.[30] Lawrence was admitted as an elector in Windsor on April 1, 1839. He appears in Connecticut River Company pay records each year from 1830 to 1833.

John Morron *(Moron, Moran)* owned property on the bank of the Connecticut River next north of the Griswold property. His property included the whole of Moran's Basin *(the Upper Basin)*. The construction of the canal divided this land leaving a two-acre parcel between the canal and the river, and a larger parcel west of the canal. On October 1, 1835, Lawrence McMahon purchased from Morron the piece between the canal and river.[32] McMahon placed a modest home on this parcel shortly after

---

[29] Hale Index of Connecticut Gravestones, https://www.halecollection.com.

[30] National Library of Ireland, Catholic Parish Registers, https://registers.nli.ie/parishes/0896

his purchase, where he raised his young family and began small farming *(which included "a ploughed field,"* [33] *fruit trees, and livestock)*. Although the available pay records do not extend beyond 1833, it is likely Lawrence McMahon continued to work for the Connecticut River Company. It is possible that McMahon built the home on the canal bank, but it is more likely that he purchased and moved to the site one of the buildings from the former labor camp. On September 2, 1841, he entered into a perpetual lease with John Morron to draw fresh water from a spring on. Morron's remaining property and carry it under the canal via lead pipe.[34]

Following Lawrence's death in 1844, Johanna McCarthy McMahon continued to live in this modest home on the canal bank until just prior to her own death in 1860. The children of Lawrence and Johanna McMahon sold the property on the canal bank to Michael Gilligan on April 19, 1869.[35]

Several children of Lawrence and Johanna McMahon survived to adulthood and raised families of their own. The third child, Johanna, married a moulder and factory worker named James Finley. They moved to Missouri where they raised seven children. The fifth child, Margaret, married Patrick Quinn, who worked as an engineer in one of the canal bank paper mills. They raised seven children in Windsor Locks, including Patrick "Paddy" Quinn, who attained a level of infamy in the pages of *The Windsor Locks Journal* for his rowdy antics in the streets and barrooms of the town. The sixth child, Mary, married a musician named Patrick Ross. They moved to Cohoes, New York where they raised four children. The seventh child, Ellen, married paper mill worker James Outerson. They raised two boys in Windsor Locks. Ellen McMahon Outerson died in Windsor Locks on August 21, 1912.[36]

There is no record that either male child or the other two female children of Lawrence and Johanna McMahon survived to adulthood.

---

[31] Connecticut River Company Records, Connecticut Historical Society, Edgar F. Waterman Research Center.

[32] Deed from Morran to McMahon dated October 1, 1835, recorded August 5, 1836 in Vol. 35, P. 1 of the Windsor, Connecticut Land Records.

[33] Deed from Estate of Gilligan to H. R. Coffin dated June 12, 1872 and recorded June 16, 1872 in the Windsor Locks, Connecticut Land Records.

[34] Lease, Morran to McMahon dated September 2, 1841, recorded February 14, 1842 in the Windsor, Connecticut Land Records.

[35] Deed from the heirs of Lawrence McMahon to Michael Gilligan dated May 3, 1869 and recorded June 3, 1869 in the Windsor Locks, Connecticut Land Records.

Michael Moore

Michael Moore appears as a laborer in the Connecticut River Company records in November 1831.[37] Further records on Michael Moore have not been found, but on April 18, 1846, a person named John Moore purchased from the Connecticut River Company two 40 foot wide by 140 foot deep lots on the south side of Grove Street.[38] During the course of construction of the canal, the company acquired several large parcels of land along the riverfront. In the 1840's, the company subdivided the land it owned to the west of Main Street and to the east of Chestnut Street and began selling off the lots. The two lots sold to John Moore were among the earliest sales and the eastern boundary is just 198 feet from the west line of Main Street. Moore paid $200 for the lots. Census records indicate that John Moore was born in Ireland around 1810. No records have yet been located establishing a connection between Michael Moore and John Moore.

Edmund Burke

The name Burke is also one of the Irish family names mentioned in the Coogan article. Edmund Burke appears in the 1830 U.S. Census for Windsor. He was living with his spouse and two sons under the age of five in a shanty in the canal laborer camp. He appeared in the Connecticut River Company pay records for July and December of 1830, and in September 1833.[39] John Burke, born in Ireland about 1824 is listed as living in Windsor Locks in the 1860 U. S. Census,

---

36 Ancestry.com

37 Connecticut River Company Records, Connecticut Historical Society, Edgar F. Waterman Research Center.

38 Connecticut River Company to Moore, April 18, 1846, Windsor, Connecticut Land Records.

39 Connecticut River Company Records, Connecticut Historical Society, Edgar F. Waterman Research Center.

but no records have yet been located establishing a connection between him and Edmund Burke.

Michael FitzGerald

The name FitzGerald is also one of the Irish family names mentioned in the Coogan article. Michael FitzGerald appears in the 1830 U.S. Census for Windsor. He was living with his spouse and one son under the age of five in a shanty in the canal laborer camp. He appeared frequently in the Connecticut River Company pay records for 1830, but not beyond then.[40] Michael FitzGerald and Mary FitzGerald appear in the baptismal register of Rev. R. D. Woodley, who on July 10, 1829, baptized their four-week-old son John during one of his three visits to the canal. This is an important record for several reasons. Not only does it identify the names of Michael, Mary and son John, but his case is one of the rare instances where a person who is identified in records prior to the November 29, 1829 ceremonial canal opening also appears in records identifying workers on site in 1830 and beyond.

John, Joseph, and Daniel Hayes

John Hayes, Joseph Hayes and Daniel Hayes appear in the Connecticut River Company records as laborers working for contractor Abiel King in 1830.[41] John Hayes is also listed as living in Suffield with a spouse and daughter under the age of five in the 1830 U. S. Census. Surprisingly, John Hayes was often working with a team and therefore paid at the rate of $2.00 per day. He also was paid $1.00 by King for the "use of shovels 40 days." If John Hayes was an Irish immigrant laborer, it would have been unusual for him to accumulate these assets in such a short time. A John and Mary Hayes appear in the baptismal register of Rev. R. D. Woodley, who performed the baptism of their four-month-old daughter Margaret on May 20, 1829. James Hayes served as a witness for a September 1828 baptism and Mary Hayes for a July 10, 1829 baptism. A James Hayes is also identified as one of the workers enticed away from the job in the

---

[40] Ibid.

[41] Ibid.

1828 trespass lawsuit by contractors Elisha Punderson and Lyman Atwater against several Suffield, Connecticut men.

Martin Doyle

Available records confirm that Martin Doyle worked regularly on the Enfield Falls Canal between the beginning of 1830 and end of 1833.[42] It is possible he worked prior to 1830 and likely he worked beyond 1833. He is listed in the 1830 U. S. Census living in Suffield near the canal with his spouse and two daughters under the age of five years old. He was not depicted as a landowner on the 1827 Canvass White map. For him to be considered a head of household in 1830, he and his family would have to be living independently in a home. He is listed in the 1830 census just prior to James Ives, which would place the Doyle home near the site that has been suggested as the location of the Suffield laborer camp. It is possible that Martin Doyle and his family were living in a shanty owned by the Connecticut River Company within that camp. The most probable location for this camp was on the two parcels of land that the Connecticut River Company acquired from Dan King and Tabitha Simond in August of 1828.[43] These properties were just north of the point at which Stony Brook empties into the river.

Research by Wendy Taylor, a librarian at Suffield's Kent Memorial Library, indicates that Martin Doyle was married to Eleanor "Ellen" Belcher and that together they reared six daughters, Margaret, Nancy, Lucy, Jane, Elizabeth, and Helen. Martin Doyle died of consumption about 1838 and is buried in Suffield's Old Center Cemetery. After Martin Doyle's death, his widow married John Mead and had two additional children. The baptismal records of Rev. R. D. Woodley include an entry for a baptism of a one-month-old child, Ann Doyle, to parents Ausitn and Ellen Doyle. This ceremony took place during his visit to the canal in September 1828. It is likely that Reverend Woodley erred in recording the name Martin as Austin, and that the child Ann Doyle is Nancy Doyle, as the birth dates align, and the name Nancy is a derivative of Ann.

---

[42] Ibid.

[43] Warranty Deed from Dan King to the Connecticut River Company dated August 14, 1828 and recorded September 30, 1828 in Vol. 17, p. 104 of the Suffield Land Records; Warranty Deed for Tabitha Simond to the Connecticut River Company dated August 14, 1828 and recorded September 30, 1828 in Vol. 17, p. 105 of the Suffield Land Records.

CHAPTER NINE

Completing the Canal

The Flood of 1828

There had always been a fear that the Spring freshets in the river caused by the melting of snow and ice upriver would prove to be more than any man-made improvements could handle. Canvass White designed the canal with such periodic floods in mind. But it was not a Spring freshet, but flooding of the river in the late Summer of 1828, that hampered efforts to complete the works at the Enfield Falls.

Heavy rains on Wednesday September 3rd and Thursday September 4th caused the water level in the river to rise at the rate of more than one foot per hour. The water level continued to rise, although less rapidly, until Sunday September 7th when it crested at twenty-two feet above the low water mark.[1] Hartford newspapers acknowledged that timber from the worksite of the Enfield Falls Canal had been seen floating down the river, but the early reports from Hartford focused more on the tremendous damage to the Farmington Canal and minimized the damage to the Enfield Falls canal.[2] New Haven papers did just the opposite.

---

[1] *The Hartford Courant*, September 9, 1828.

[2] Ibid.

The New Haven Register reported almost immediately that "the canal and locks constructing near the Enfield Falls, by the Hartford Company, have suffered immense damage" and that "damage on the Farmington Canal is trifling compared with the losses sustained by the friends of river improvements."[3] *The New Haven Herald* reported: "We also learn that the works at Enfield Falls are mostly swept away. In attempting to save some timber, two men *(Irish)* were swept over the falls, one of whom was drowned."[4]

Commentator Nelle Grace Abbe reported, nearly 69 years later, "One of our oldest inhabitants affirms that about the first thing he recollects was hundreds of wheelbarrows, used by workmen in building the canal, floating off down the river in a big freshet."[5]

Similarly, Jabez Haskell Hayden reported, "during the work a freshet in the river carried away all that had been done at that point, and above.[6]

There is no need for conjecture about the amount of completed work that was washed away in the September 1828 flood. Within the Canvass White papers is a document dated November 1, 1828, which is a section by section recapitulation of the amount of work completed and remaining to be done on the canal.[7] The document also includes White's calculations as to the amount of completed work on each Section that was washed away by the flood. The work is measured in cubic yards for rock excavation, feet for the wing dam and pier, cubic yards for embankment, cubic feet for wharfing, and perch of stone for slope walls, locks, aqueduct and other masonry. White converted each of these quantities into dollars to calculate the value of the work performed to date, the anticipated cost of the remaining work, and value of completed work lost in the flood.

There was no accounting for the loss of wheelbarrows and lumber that were not recovered. As demonstrated in the following

---

3 Reports of *The New Haven Register*, as cited in *The Hartford Times*, September 16, 1828.

4 Report of *The New Haven Herald*, as cited in *The New York Evening Post*, September 10, 1828.

5 Nelle Grace Abbe, "Traffic on the Connecticut River Half a Century Ago", *Connecticut Quarterly*, George Atwell, Hartford 1897.

6 Hayden, *Historical Sketches*, Windsor Locks Journal Publishing Company, Windsor Locks, 1900.

7 Estimate of Work Done and Remaining to be Done on the Enfield Falls Canal, November 1, 1828. Canvass White Papers, Rare and Manuscript Collection, Carl A. Kroch Library, Cornell University Library.

chart, White summarized his calculations:

Section	Estimate of Work Done	Estimate of Work Remaining	Carried Away by Flood
One	$13,299	$7,085	$800
Two	$10,190	$11,289	$4,221
Three	$1,878	$16,733	$1,303
Four	$7,810	$2,956	$558
Five	$8,367	$433	$882
Six	$8,087	$5,206	$50
Seven	$7,370	$7,836	$1,268
Eight	$2,400	$277	0
Nine	$3,797	$866	
Ten & Eleven	$1,767	$6,486	0
Guard Lock	$5,456	0	0
Lift Lock One	$6,347	0	0
Lift Lock Two	$6,051	0	0
Lift Lock Three	$6,863	0	
Aqueduct	$5,999	0	0
Culvert	$1,739	0	0
Waste Weir	$1,500	0	0
3 Swing Bridges	$1,500	0	0
Roads	$278	0	0
Moving Buildings	$1,000	0	0
Total	$78,015	$85,411	$7,681

Surely, the flood caused a frustrating delay in the time it would take to complete the Enfield Falls Canal. Athough disappointing to the Directors, the flood did not result in a devastating loss to the Connecticut River Company and its stockholders. The rejoicing by the New Haven Canallers over the effects of the flood upon the progress of the works at Enfield Falls proved to be unwarranted.

Building Steam-Powered Tow Vessels Capable of Passing Through the Locks of the Canal

Although the Enfield Falls Canal was designed with a traditional towpath to enable draft animals to tow vessels through the canal, the Directors of the Connecticut River Company were banking on a fleet of steam-powered tugs to tow barges laden with goods through the length of the canal. This innovation would set the project apart from its competitors. These aspirations were limited by the fact that such vessels did not yet exist.

The towboats would need to be of shallow draft, short enough, and narrow enough to fit within the canal's lock chamber.

Luckily, noted inventor Thomas Blanchard of Springfield, Massachusetts, was a strong proponent of the proposed improvements to upriver navigation, and turned his attention to the building of the steamships needed to support that cause. Blanchard had invented two machines to streamline the production of guns at the nation's two national armories at Springfield and Harpers Ferry. The first was a machine that would uniformly cut the exterior surfaces of musket barrels. The second was the "Blanchard lathe" which was initially used to mechanically produce gunstocks and subsequently adapted to automate the production of shoe lasts, axe handles, wagon wheel spokes, and dozens more irregularly shaped wooden products.[8]

Thomas Blanchard, developer of the sternwheeler Vermont.

In 1826, Thomas Blanchard built the first American automobile, a 2,000 pound steam-powered machine that he drove in Springfield.[9] Although Blanchard's "horseless carriage" was not a commercial success, it endowed him with a thorough knowledge of the design and operation of steam engines, which he relied upon to develop two shallow draft steam vessels that could be used on the upper Connecticut River and other rivers.

Blanchard's first steam vessel, appropriately named the *Blanchard,* was designed and built during the first half of 1828 and began in water trials in late July and early August of that year.[10] The *Blanchard* was a side-wheel steamer, and as such was not particularly well suited for navigation through narrow canal locks. The following year, Blanchard began

---

[8] National Park Service, Springfield Armory, Thomas Blanchard, https://www.nps.gov/spar/learn/history-culture/thomas-blanchard-and-his-lathe.html.

[9] Thomas Blanchard, The American Society of Mechanical Engineers, https://www.asme.org/topics-resources/content/thomas-blanchard.

[10] *Springfield Republican*, August 6, 1828.

river trials with his second steamer, called the *Vermont*. The *Vermont* was a sternwheeler and well-suited for canal use. Blanchard determined that the *Vermont* could gain considerably more power by moving the stern wheel farther back from the stern compared to its predecessor stern-wheeler, the promotional *Barnet*.[11] The July 14, 1829 *The Hartford Courant* reported:

> The Steam Boat *Vermont*, with fifty passengers, came to our wharf yesterday in two hours and fifty minutes from Springfield, including two stops. She returned in the afternoon. On Saturday last she came down and ascended Enfield Falls twice – once in an hour and twenty-five minutes, and the other time in one hour and eighteen minutes, by the power of her steam only. The distance up the falls is about five miles. The boat is driven by one wheel, with a double set of buckets, and placed in the stern. The engine, hull, etc. were all built and fitted under the immediate direction of Mr. Blanchard, one of the most enterprising and ingenious mechanics in our country, who seems determined that if the river shall not be adapted to steam boats on the old plan, he will adapt the boats to the present state of the river.[12]

The *Vermont* made an excursion to Brattleboro, Vermont, on August 5, 1829. The next day she travelled to and passed the locks at Bellows Falls and then made it as far as the Quechee Falls, where the locks were too small to permit further passage north.[13] The *Vermont* was eighty feet long, fourteen feet wide, and drew only 15 inches of water.[14] She could travel at six miles per hour upstream, a speed unattained by its predecessors.[15]

The End in Sight

Although not without setbacks and delays, progress on the Canal at Enfield Falls continued to be made throughout the remainder of 1828 and 1829. On August 7, 1829, *The Hartford Courant* reprinted an editorial that first appeared in *The Bellows Falls Examiner* on August 3, 1829. The author of the editorial is identified only by the letter "W." It read:

---

11 *The Hartford Courant*, July 14, 1829.

13 *Bennington Vermont Gazette*, August 11, 1829, reprinted from *The Brattleboro Messenger*, August 10, 1829; "The Navigation of the Connecticut River," *Proceedings of the Vermont Historical Society For the Years 1915-1916*, p. 68, Vermont Historical Society, 1918.

14 Ibid.

15 *The Hartford Courant*, Navigation of the Connecticut River, August 14, 1868.

Mr. Editor – As much anxiety is felt by the people in the central and northern part of the Connecticut River Valley, respecting the contemplated improvement in the navigation of the river, and as all are anxious to know what progress is making, it must be gratifying to the friends of internal improvement to learn that there is a fair prospect that the canal, by Enfield Falls, will be completed the present season, as the work is already in great forwardness – and when completed will be immensely important to the public, as the passage up over Enfield Falls is one of the most difficult and dangerous to navigate of any portion of the River between the city of Hartford and Barnet. This canal is five and a half miles in length, and is of sufficient width and depth, to admit vessels passing, that carry fifty tons or more, if necessary. At the head of the falls, or upper end of the canal, there is a wing dam and guard lock, now building, of the best materials, and the embankment for about four miles below is much of the way entirely finished, and is for a great distance more than twenty-one feet high and sufficiently wide upon the top for a good tow path, or even a horse and chaise to pass safely. For nearly four miles the canal is between this embankment and the natural bank of the river, and forms what, in any other than the present age, would be called a most stupendous work. From the guard lock to the lift locks, the canal is perfectly level, being a distance of more than five and one fourth miles. Here the water from the canal passes through three locks of ten feet lift each, into the river. These locks are ninety feet long and twenty feet wide, composed entirely of stone and laid in cement, in the most solid and beautiful manner. Between each of these locks, there is a basin two hundred feet long and wide enough to admit boats or vessels to pass up and down at the same time. The Connecticut River Company have already expended, in this vastly important work, more than one hundred and forty thousand dollars, and it is believed it will require forty thousand more, to finish the canal. But the whole so far as is completed, is done in the most permanent manner, both as regards the materials and the workmanship – and does great credit to the superintendent, the engineer and the contractors. The stock to complete the canal has all been taken up, and the public may feel assured it will be open for use the present fall, unless the river should be so high, as to prevent the contractors from prosecuting their work.[16]

As winter approached, a tremendous push was made to complete the project in the current season. On November 7, 1829, *The Connecticut Mirror* reported:

> We understand that the canal at Enfield Falls is nearly ready for navigation. The water has been let gradually upon the banks which, thus far, sustain the pressure well; and it is hoped the canal will be filled, so far as is intended this season, before Wednesday next, so that boats may pass on that day.[17]

---

16 *The Hartford Courant,* August 7, 1829.
17 *The Connecticut Mirror*, November 7, 1829, as reprinted in *The Brattleboro Messenger,* November 13, 1829.

Canal Opens

CONNECTICUT COURANT
Hartford, November 17

Canal at Enfield Falls. - The Canal recently constructed round Enfield Falls having been filled nearly to the height to which it is intended to admit the water the present season, a number of gentlemen from this city, and other places on the river, visited the works on Wednesday last, for the purposes of witnessing the first passage of boats. A party from this city arrived at the foot of the Falls in the steam-boat Blanchard, in carriages about 10 o'clock, where they were met by another party from Springfield, in the steam-boat Vermont, which had recently returned from her excursion to Windsor, in Vermont. In her passage down from Springfield, she passed through the whole length of the canal, and came into the Connecticut below the Falls. The stockholders present, with other gentlemen from Hartford, Springfield, and the neighboring towns, then went aboard the Vermont, and two other boats towed by horses, and set sail for the head of the Falls. The boats were an hour and ten minutes in passing through the canal, a distance of five and a half miles, including the detention at the locks. At this place, after an exchange of friendly salutations, the gentlemen from Springfield parted from the company, and proceeded on the passage home in the Vermont. After a short time spent in examining the works, and particularly the excellent and substantial construction of the Guard Lock, the rest of the party returned in the boats down the canal to the foot of the Falls. The whole time occupied in passing up the canal and returning, including the delay at the head of the Falls, was two hours and a half. The width of the canal is so ample that the steam-boat, having her wheel in the stern, and moving at the rate of five miles per hour, caused but little commotion in the water, and not the slightest injury to the embankment. Sixteen boats, loaded with merchandise, passed through the canal the same day.

The firm and substantial manner in which the locks and embankments are constructed, reflecting as it does much credit on the contractors, and the agent under whose superintendence the work has been prosecuted, was much admired by the gentlemen present. It is almost superfluous to add that the excursion was attended with a high degree of interest, and the party returned home much gratified with the scene they had witnessed.

The successful termination of an undertaking so important to a free navigation of the river, is a just cause for hearty congratulation, with all who wish to see an easy communication opened between this place and the towns in the valley above us. This improvement has not been effected without considerable expense, but the importance and utility, we believe, will soon be demonstrated by the increased facilities it will afford to the trade of the valley – while at the same time, it is hoped, the public spirit of those who promptly advanced the funds for its accomplishment, will not loose its reward.

Beside the more immediate and direct advantages which may reasonably be anticipated from the construction of these works, there is another point of view in which they must be regarded as highly important – they furnish the most satisfactory evidence of what may be accomplished in the improvement of river navigation. They show, by actual experiment, what some have been disposed to question, that the pools in the Connecticut may be so connected by short canals, as to afford a regular and valuable steam-boat navigation. By means of the canal at Enfield, which is less than six miles in length, steam-boats may now ply, without interruption, a distance of nearly 40 miles. Similar works, of less magnitude, on other sections of the river, where the navigation is at present obstructed, would, without doubt, be attended with corresponding results. We subjoin a brief description of the works at Enfield Falls.

The canal commences at the head of the Falls, by a wing dam 700 feet long, which reaches to the middle of the river. From the lower end of the wing dam a pier extends down 200 feet parallel to and 100 feet from the west bank, and is raised above the river so as to form a basin and safe entrance to the Guard Lock. At right angles to this pier, a breast wall of solid masonry, strengthened by buttresses, extends 70 feet towards the bank, and is there united to the Guard Lock. This and the breast wall are 16 feet higher than the surface of the water in the canal, presenting a firm defense against the highest floods. The breast wall covers twelve sluices, with sliding gates, for the free admission of water for hydraulic purposes. The river banks are generally high and rocky, for about three miles below the Guard Lock, and the canal is formed by an embankment of earth raised in the bed of the river, and protected on the outside by a stone wall. Two miles below the Guard Lock, Mill Brook crossed the line of the canal, and is passed by an aqueduct of 90 feet long and 60 feet wide, having six piers and abutments of substantial masonry. The height of the artificial embankment increases gradually, as it passes down the river, till is rises to a perpendicular elevation of 25 feet, when the high river bank retires to the west, and the canal is carried about two mile over land to its termination below the Falls. Here are three locks of masonry of ten feet lift each. The locks are separated by pools fifty feet wide, in which ascending and descending boats may pass each other, and avoid the detention which is unavoidable where several locks are combined in a connected line. The dimensions of the locks are 90 feet by 20 feet in the clear, and they are calculated to have four feet depth of water. The depth of the canal varies from four to twenty feet. The average width at the surface of the water, is about seventy feet, and the total length five and a half miles.[18]

Equally enthusiastic reports regarding the opening of the canal appeared not only in the newspapers of the Connecticut River Valley towns, but in newspapers as far away as Lancaster, Pennsylvania.[19]

---

18 *The Hartford Courant*, November 17, 1829.

19 *Lancaster Intelligencer*, December 4, 1829.

CHAPTER TEN

Beyond Completion

Both Canvass White and Alfred Smith were under tremendous pressure to open the Enfield Falls Canal before the close of the 1829 building season. While its ceremonial opening on November 29, 1829 must have been a great relief to all concerned, significant construction work remained. Additionally, the company had neither selected a person to manage the canal, nor developed procedures to operate it. The Company had just a few short months to make the canal ready for use as soon as the ice cleared in the Spring.

There are no documents in the Canvass White papers that shed light on the construction activities on the canal following the detailed recapitulation of the work completed and estimate of work remaining prepared by White on November 1, 1828. The information available for 1829 is scant, except for the many newspaper accounts of the ceremonial opening in November. The Connecticut Historical Society, with a great deal of foresight and good judgment, has preserved some records of the Connecticut River Company in its archives. Within this collection are receipts for payments made by the Company between 1830 and 1833.[1] These include payments to several contractors engaged to complete the construction of the canal after its ceremonial opening. The records also detail payments made to local individuals. These men

---
[1] Connecticut River Company Papers, Connecticut Historical Society.

owned teams of horses or oxen and were hired to assist with the final construction or to help repair leaks. Manual labor was provided by both local citizens and Irishmen to complete the canal. During the first year of operation, while the rush to complete the canal was still at full throttle, these individual laborers continued to work for contractors. In subsequent years, a small number of these laborers remained on site, and were engaged directly by the Connecticut River Company to perform ongoing maintenance and repair.

Contractors

Abiel King

Abiel King lived on farmland he owned on the canal in Suffield north of King's Island.[2] Through the winter of 1829 until November of 1830, King worked as a contractor for the Connecticut River Company.[3] It is possible he performed work for the Company prior to that time, but no records confirming his earlier service have yet been discovered. In November of 1830, King submitted a summary of the work he and his subcontractors performed during the 1830 building season. The summary begins with a charge of $2.50 "to care for the canal through the winter." The summary is not specific as to the portion of the canal on which he was contracted to work, although he makes several references to a float bridge. The float bridge was at a point on the canal north of the aqueduct.

King paid his subcontractors at various rates ranging from $.65 cents per day to $.80 per day for unskilled work, and $2.00 per day for work with teams of horses or oxen. He billed the Company $1.25 per day for his own work. The records include pay receipts to contractors who worked and were paid directly by King, as well as some of King's weekly invoices for reimbursement from the Connecticut River Company. Among the subcontractors who worked regularly for King were brothers Joseph Hayes, John Hayes, and Daniel Hayes, who were paid $.65 per day. John Hayes also occasionally worked with a team at the rate of $2.00 per day. Other subcontractors who worked regularly

---

[2] Canvass White map of the proposed canal, June 20. 1827; *United States Census Records 1820* (1830).

[3] Connecticut River Company Papers, Connecticut Historical Society.

for Abiel King were brothers John McMahon and Lawrence McMahon, Michael FitzGerald, Martin Doyle, Michael Guinane, Patrick Rooney, Michael Murphy, Edmund Burke, Thomas Ryan, and Thomas Morris. All of these men were Irish immigrants. They were paid from $.65 to $.75 per day for their work. Others who worked for King, but who do not appear in multiple pay records, are Frank Gilday, Patrick Timmons, Daniel O'Connell, John O'Neill, Daniel Deland, Michael Halley, Timothy Cain *(Kane)*, Michael Connery, Nathaniel Gates, Hugh Murphy, Felix Maguire, James Kennan, William Towers, John Donoho, Abraham Stewart, John Maloney, and Edward Ryan. These surnames suggest that these men were also Irish immigrants. Some local residents were also engaged by Abiel King, including Orrin Parsons, Quartus Parsons, Charles Abbe, Jr. *(the son of Charles Abbe),* Cyrus Pease, Munroe Briggs *(the son of Greene Briggs)*, Levi Pease, and James Allen *(the "hired man" of James Ives)*. These men *(or their fathers in the case of Abbe and Briggs or their employer in the case of Allen)* were paid $.75 to $.84 per day.

Abiel King hired Peter Parker to assist with making and repairing bridges, and many of the entries in his summary concern bridge work performed by Parker.

Patrick Kalter *(Coulter)*

Patrick Kalter[4] was engaged as a contractor by the Connecticut River Company from March 1830 through August 1830.[5] He submitted two invoices for his work, both of which survive in the Connecticut Historical Society archives. The first invoice included all work between March 18, 1830 and June 26, 1830. Some of the work itemized in the first invoiced included, "work near the Fish House," "digging and placing timber in the Basin," "levelling the embankment on the west side of the basin at the locks," "digging ditch and puddling back of the Fish House," delivering a "boat load of earth to the culvert," "work in the water at the Guard Lock," "clearing away part of Ives' bridge," "night watching on the embankment," "stopping leak near Lamberton's House," "work on the road," "work at the Aqueduct," "placing timber from the canal at into the

---

4   The name is possibly a phonetic spelling of the Irish name Coulter.

5   Connecticut River Company Papers, Connecticut Historical Society.

basin near Widow Griswold's," "levelling the embankment on the east side of the Basin at the locks," and, "work at the waste weir." A second invoice, dated August 3, 1830, was for "work done on the road opposite the upper lock at the foot of the canal."

Much of the work identified by Kalter was work that by necessity needed to be completed near the end of the project. For example, the design called for two small basins situated between the three lower lift locks so that boats passing up and down the locks could pass each other. These basins could not be formed until the locks were complete. The upper (Morran's) basin and lower *(Carleton's)* basin could not be finished and filled until the canal was complete.

The work by the back of the "Fish House" refers to the canal as it passed behind the home of Corning Fish which was on the riverbank. The Fish property is at the point where the canal transitions from a channel created by making an embankment in the river to a channel formed by the excavation of dry land. The canal makes a sharp turn to the west at this transition point such that it passed behind the Corning Fish house on the bank of the river at what was known as "Fish's Meadow." Puddling is the process of applying clay to the bottom and sides of the canal prism to prevent water from escaping. Ives' Bridge is a reference to the float bridge which was at a point on the canal, north of the aqueduct, abutting the land of James Ives. This bridge was located at an ideal location to facilitate the crossing of workers from the Suffield labor camp on the west side of the canal to ongoing construction sites on the east side of the canal.

The road running north from Windsor to Suffield needed to be relocated to the west at several points to make way for the canal. The location of the existing road and the anticipated location of the road, post construction, is depicted clearly in Canvass White's 1827 map. This road relocation and finishing work is referred to in both of Kalter's invoices.

The place of Patrick Kalter's birth has not been definitively determined, but the first name Patrick and his use of the symbol "X" to sign his name on the receipts suggests he was an Irish immigrant. Restrictions on the education of Catholics in Ireland meant that the majority of canal era Irish immigrants were unable to read or write the English language. Although the surname Kalter is of German and Jewish origin, it is possible that since he was unable to sign his name, it was a phonetic misspelling of the Irish surname Coulter.

Ambrose Davis

Ambrose Davis was a young mason who in 1830 was living with his wife Cynithia and one young son near the canal in Windsor. He submitted two invoices to the Connecticut River Company for work performed on the canal. The first was for "sundry jobs" on the canal from April 10, 1830 to June 22, 1830. The second was for "laying 38 perch of stonework on the canal Section No. 7," which was submitted and paid on December 2, 1830. The specifications for Section 7 included excavation, embankment, and wharfing and it includes the area near Fish's Meadow where the canal enters the Corning Fish property.

Orren Bissell

Orren Bissell was a millwright from East Windsor, Connecticut. He was engaged by the Connecticut River Company in May of 1833 to perform repair work on canal bridges and to reinforce the existing bridges and waste weir gates with wrought iron.

Other Local Contractors

Much of the remaining work was completed by farmers from Suffield and Windsor living adjacent to or nearby the canal. As farmers, these men had the beasts of burden and tools needed to perform the plowing and scraping needed to move large quantities of earth. The landowners who also served as canal contractors include Daniel Lester, Enos Morron, Harris Haskell, Williston Griswold, Gaylord Griswold, Talcott Mather, Newton Smith, and Zebrina Bronson.

The Lock Tender

The duties of the Collector of Tolls, or Lock Tender, were far greater than passing boats through the canal and collecting tolls. The Lock Tender was the on-site superintendent of the canal. He managed all maintenance issues, repairs, and improvements to the physical plant. He handled all financial tasks which, by their nature, had to be performed locally, including the payment of employees and contractors. He was also the public face of the canal to all travelers, business partners, and local citizens.

The position demanded a person of character and reliability. The Board of Directors appointed Asa Barr Woods, who was the Toll-Gatherer at the bridge across the Connecticut River at Hartford at the time,[6] as the first Lock Tender of the canal in early 1830.[7] Woods was born March 28, 1794, in Ware, Massachusetts. He was living in Belchertown, Massachusetts at the time of his marriage to his first wife Panthea Pease of Somers, Connecticut on June 19, 1817. With Panthea Pease, Woods had a son, William B. Woods, who became a prominent physician in Somers, Connecticut. Following his first wife's death on June 16, 1821, he married his second wife, Elizabeth House, of Clinton, New York. With Eliza, he had one daughter, Harriet Woods born on August 13, 1832 in Windsor Locks. Asa Woods was a professor of religion and was among the founders of the Windsor Locks Congregational Church, were he served as a deacon, and where both he and his wife were instrumental in establishing a Sunday School.[8] The fervor of A. B. Woods' religious beliefs posed a practical problem for the Connecticut River Company. Woods was unwilling to let boats pass through the canal on the Sabbath.[9] Ultimately, a compromise was reached allowing boats to pass on Sunday prior to 8:00 am and in the event of an emergency. A. B. Woods served as Lock Tender until his death on December 17, 1854 at the age of 60.[10] He is buried in Grove Cemetery.[11]

An inventory of the assets of A. B. Woods' estate reveal that he owned 56 shares of stock of the Connecticut River Company valued at $1,260 at the time of his death. He also held promissory notes in his favor from several mill owners along the canal. His receivables included

---

[6] *The Hartford Courant*, George Goodwin v. Asa B. Woods, November 12, 1822.

[7] Erving, Henry W , *The Connecticut River Banking Company 1825 – 192* (Hartford, The Connecticut River Banking Company, 1925), p.129.

[8] Hayden, *Historical Sketches,*, p. 98.

[9] Erving, *The Connecticut River Banking Company 1825 – 1925*, (Hartford, The Connecticut River Banking Company, 1925), p. 129.

[10] United States Census, 1850; Inventory of the Estate of Asa B. Woods, Hartford Probate Court Records, Probate Date December 27, 1854.

[11] Connecticut, U.S., Deaths and Burials Index, 1650-1934.

a portion of his salary not yet paid by the Connecticut River Company.[12]

Assistant Lock Tenders, Bridge Tenders, and Watchmen

Woods also engaged local citizens to serve as assistant lock tenders, bridge tenders, and night watchmen. Connecticut River Company records from the first three years of operation[13] note the service of the following individuals:

Samuel Norris – Guard Lock Tender
1830 April to Sept 20, 1830. Paid $60.00 on Oct. 29, 1830.

Elizur Lamberton – Bridge Tender
For Raising Canal Bridge for Steam Boats Vermont & Spring field 96 times @ 6 cents. Paid $5.75 on Dec 31, 1830

James Pomeroy – Watching Canal and Lower Locks Tender
April 1 to Dec. 22, 1830 assisting in tending locks & small jobs on gates. Paid $93.75.
March 22, 1831 to May 24, 1831 tending locks @$1. Paid $22.
July 4, 1831 to July 31, 1831. 29 nights watching canal at $1.00. Paid $29.00.
Aug. 1 to Nov. 25, 1831. 97 nights watching canal at $.75. Paid $72.75.

Marcellus Pinney - Lower Locks Tender
7 ¾ months ending December 17, 1832 @$10. Paid $70.75.

Newton Smith – Guard Lock Tender
For tending Guard Locks one year ending Dec. 26, 1932. Paid $50.
For tending Guard Lock one year ending Dec. 31, 1833. Paid $65.

---

[12] Inventory of the Estate of Asa B. Woods, Hartford Probate Court Records, Probate Date December 27, 1854.

[13] Connecticut River Company Records, Connecticut Historical Society, Edgar F. Waterman Research Center.

Daniel Lester – Guard Lock Tender
No dates listed. For tending guard locks. $30 paid April 29, 1831.

The Samuel Norris who served as the lock tender on the guard lock from April to December in 1830, during the canal's first season of operation, is the same man who allowed Reverend Bernard O'Cavanaugh to stay in his home and celebrate mass for canal workers in his parlor during his September 1829 visit to the area.

Elizur Lamberton is listed in the 1820 and 1830 U.S. Census living in Windsor near the southern end of the canal. Jabez Haskell Hayden. who lived on South Main Street, refers to a Lamberton family living near him.[14] Windsor Congregational Church records show that Lamberton married Mary Winslow on April 29, 1811, and that their two-year-old son Elizur, Jr. tragically died by scalding on November 19, 1813.[15]

James Austin Pomeroy was born in Suffield, Connecticut in 1801. His parents were Jonathan Pomeroy and Thersa "Terza" King Pomeroy. He died in 1840 and is buried in West Suffield Cemetery.[16]

Marcellus Pinney was born May 6, 1812 to Oliver Pinney and Lois Pease Pinney. He subsequently moved to Springfield, Massachusetts where he married Amelia Robinson on December 14, 1836. His occupation is listed as farmer on the 1850 United States census and he was working at the United States Armory in Springfield at the time of the 1860 Census. He also acquired a great deal of government land in Iowa, although he did not move there. He died in Springfield May 6, 1897. His wife Lois Pease Pinney died March 2, 1893. They are buried in Springfield Cemetery.[17]

Newton Smith was born in Suffield in 1804 to Elihu and Lydia Smith. He married a woman named Ruth who was born about 1807 in Vermont. Ruth's maiden name is unknown. His occupation was listed as a traveling merchant or peddlar in the Middlesex County, Massachusetts area in the U. S. Census for 1850, 1860, and 1870. He died July 12, 1870

---

[14] Hayden, *Historical Sketches*, p.98.

[15] Ancestry.com

[16] Ibid.

[17] Ibid.

in Lowell, Massachusetts.[18]

Daniel Lester was the son of Captain Daniel Lester and Abiah Lester. All are buried in the Old Center Cemetery in Suffield.

Buildings

The Connecticut River Company contracted with several local builders to construct the buildings necessary for its operations. On September 13, 1830, the Company paid Suffield Builder Zephidiah Sherman $343.00 for "erecting a dwelling house near the Guard Lock on the canal and furnishing materials." This building was a dwelling for the Lock Tender, who at the time was Colonel Samuel Norris. A subsequent deed from Norris to the Connecticut River Company confirms the location of this home.[19] It was on a seven rods wide *(115.5 foot)* lot at the eastern end of the Norris' property, bounded to the north by the widow and heirs of Amos Kendall, and to the east by the Connecticut River. The length of the lot was continued south to the point that the area of the lot would equal one-half of an acre. The property was bounded on the west and south by a recently built fence. The conveyance included the right to pass onto Norris's land to make repairs around the Guard Lock. A curious feature of the conveyance is that it contained a restrictive covenant that, "said Company shall not use the house lot or any house that shall be built by them on said lot as a tavern." Title to the lot reverted to Samuel Norris or his heirs or assigns if the restriction was ignored.

In June of 1831, the Connecticut River Company paid Pomeroy & Lester $5.00 for "building a Shantee as per agreement." No location for the shanty is stated but it was probably built on the site of the Suffield labor camp, about one-half mile north of the aqueduct, which had been purchased from Tabitha Simonds and Dan King in August, 1828. It is unclear why a shanty would need to be built as construction was coming to a close. The most likely explanation is that it served as a shelter or cabin for canal watchmen. This work may have been connected with the work at Ive's Bridge, which was the float bridge crossing the canal near the Suffield labor camp.

---

[18] Ibid.

[19] Deed from Samuel Norris to Connecticut River Company, October 26, 1835, recorded October 26, 1835, Vol. 19, p. 208, Suffield, CT Land Records.

In September 1831, the Connecticut River Company paid Otis Pierce $15.00 for framing a barn and in October 1831 paid $13.12 for "covering barn, building wood shed, and making a fence by here." The location of the work as noted on the receipt is Suffield. This work was to construct the warming barn for the tow-horses and an adjacent woodshed built on top of the head gate and depicted in several photographs and drawings, as well as the fence referred to in the Norris deed.

On June 8, 1832, the Connecticut River Company paid Somers, Connecticut builder William Wardwell $144.16 to furnish timber for and frame a dwelling house. Wardwell used 4119 feet of timber to frame the building. In December of 1832, the Company paid Linus Parmele $43.75 for 43 ¾ days work on a "dwelling house." The finishing work by Parmele appears to be for the same structure framed by William Wardell. The purpose for which this dwelling house was built is unknown. The widow and heirs of Darius Parmele are shown on the 1827 Canvass White map as owning property on the river bank just north of the Suffield/Windsor boundary. The lot is on a bluff, situated next north of the Corning Fish property. Linus Parmelee and his family are listed as living in Suffield at or near the same location at the time of the 1830 U. S. Census. It is possible that this home was built for the Parmalee family by the Connecticut River Company because the works of the Company had in some way done injury to the original Parmele home.

Troubleshooting Leaks

The following records indicate an ongoing problem with leaks in the works during the first three years of operation. The Connecticut River Company records include these entries:

> December 27, 1831
> Received by the Connecticut River Company by the hand of A. B. Woods. Thirty-five Dollars for damages done to our Lower Grist Mill in consequence of the late Breach of the Enfield Falls Canal.
>
> Herlehigh Haskell
> Seth Dexter
> Harris Haskell
>
> April 25, 1833
> Rec. of Con. River Co. by hand of A. B. Woods fifteen dollars in full for damages done my land by the water breaking out of the canal in 1832 and

for earth to repair and stop leaks. $15.00

John Morron

8 June 1833
Recd. of the Connecticut River Co. by the hand of A. B. Woods. Eleven Dollars and 95 cts. in full for damage done my land by repairing the canal in 1832.

Williston Griswold
Joseph Pettibone

Augst 3, 1833

For damage done to my land and for earth to mend & repair canal. $18.00

John Morron

Oct. 1, 1833
Recd. of Conn River Co. by hand of A. B. Woods Twelve dollars in full for damage done our land at fish place by the water breaking out of canal and earth for repairing up to this date.

Williston Griswold
Joseph Pettibone

These entries confirm ongoing problems with leaks from the canal between 1830 and 1833. The fourth receipt to Griswold and Pettibone identifies the location of the leaks as the "fish place." The Griswold family operated a shad fishery from their property for many years both before and after canal construction. Joseph Pettibone had some connection to the Griswold property, perhaps as the manager of the shad fishing operation. The shad fishery on the bank of the river had been cut off from the main body of the Griswold land by the canal, but the family retained ownership of the land and the right to continue fishery operations. John Morran owned the property on the riverbank next north of the Griswold fish place, which had also been cut off from the main body of his land by the canal. This land between the canal and the river was also the location of the waste weir, which was intended to help control water levels in the canal by allowing for the release of water from the canal into the river. The precise location of the waste weir is depicted on Canvass White's 1827 proposed canal map. Other entries in A. B. Woods' ledger show expenditures paid to laborers during the same timeframe. Some of these were for the labor needed to repair the leaks. In consequence of the frequent breaks at the site, and to save the company the cost of paying for damages and repairs, the waste weir was permanently removed some years after its completion.

Transportation on the Canal

The transportation function of the canal was an immediate success. For passenger travel, steam powered passenger ships, also known as packets, passed daily on the route between Hartford and Springfield. In 1831, Thomas Blanchard followed up the steamer *Vermont* with the launching of the *Massachusetts*, which was a twin-engine steamer with more accommodations than the *Vermont*, including a passenger cabin. The *Massachusetts* was too large to pass through the canal, however, and had difficulty running in low water.[20] In the following years the passenger vessels *Agawam* and *Franklin*, both capable of passing through the canal, were added for passenger service north of Hartford.[21]

Famed English writer Charles Dickens passed through the canal aboard the steamer *Agawam*[22] as part of a trip from Springfield to Hartford on February 7, 1842. He wrote about the trip in his 1842 travelogue, *American Notes*.[23] The *Agawam* was one of the small steamers designed to pass through the canal and Dickens emphasized its small size in his recollection of the trip:

> We went on next morning, still by railroad, to Springfield. From that place to Hartford, whither we were bound, is a distance of only five-and-twenty miles, but at that time of the year the roads were so bad that the journey would probably have occupied ten or twelve hours. Fortunately, however, the winter having been unusually mild, the Connecticut River was 'open,' or, in other words, not frozen. The captain of a small steamboat was going to make his first trip for the season that day (the second February trip, I believe, within the memory of man), and only waited for us to go on board. Accordingly, we went on board, with as little delay as might be. He was as good as his word, and started directly.
>
> It certainly was not called a small steamboat without reason. I omitted to ask the question, but I should think it must have been of about half a pony power. Mr. Paap, the celebrated Dwarf, might have lived and

---

[20] *Ibid.*

[21] *Ibid.*

[22] There are varying reports as to whether the steamship was *The Massachusetts* or *The Agawam*. Since *The Massachusetts* was reportedly too large to pass through the canal, it is probable the trip was aboard *The Agawam*.

died happily in the cabin, which was fitted with common sash-windows like an ordinary dwelling-house. These windows had bright-red curtains, too, hung on slack strings across the lower panes; so that it looked like the parlour of a Lilliputian public-house, which had got afloat in a flood or some other water accident, and was drifting nobody knew where. But even in this chamber there was a rocking-chair. It would be impossible to get on anywhere, in America, without a rocking-chair. I am afraid to tell how many feet short this vessel was, or how many feet narrow: to apply the words length and width to such measurement would be a contradiction in terms. But I may state that we all kept the middle of the deck, lest the boat should unexpectedly tip over; and that the machinery, by some surprising process of condensation, worked between it and the keel: the whole forming a warm sandwich, about three feet thick.

It rained all day as I once thought it never did rain anywhere, but in the Highlands of Scotland. The river was full of floating blocks of ice, which were constantly crunching and cracking under us; and the depth of water, in the course we took to avoid the larger masses, carried down the middle of the river by the current, did not exceed a few inches. Nevertheless, we moved onward, dexterously; and being well wrapped up, bade defiance to the weather, and enjoyed the journey. The Connecticut River is a fine stream; and the banks in summer-time are, I have no doubt, beautiful; at all events, I was told so by a young lady in the cabin; and she should be a judge of beauty, if the possession of a quality includes the appreciation of it, for a more beautiful creature I never looked upon.

After two hours and a half of this odd travelling *(including a stoppage at a small town [the village of Windsor Locks]*, where we were saluted by a gun considerably bigger than our own chimney), we reached Hartford, and straightway repaired to an extremely comfortable hotel: except, as usual, in the article of bedrooms, which, in almost every place we visited, were very conducive to early rising.[24]

For commercial shipping, barges were towed both by modern steam powered vessels specifically designed to fit within the canal's narrow locks and by the traditional method using draft animals. A receipt within the Connecticut River Company records in the collection of the Connecticut Historical Society shows that during the canal's first year of operations the steamboat *Vermont* made thirty-seven passages through the Enfield Falls Canal between June 2, 1830 and August 13, 1830 and twenty-three passages through the canal between October 4, 1830 and November 12, 1830. The toll was $2 for downriver passages, $3 for upriver passages and $.50 per ton of merchandise transported. The *Vermont*

---

23 Erving, *The Connecticut River Banking Company*, p. 167.

was piloted by Thomas Blanchard the noted Springfield inventor during the June through August trips. The October through November trips were piloted by a Captain B. Phelps.[25] In 1831, the *Hampton* was launched as a freight towing vessel by John Cooley & Company. The *Ledyard* and the *William Hall* were also added as tow-boats later that season.[26]

The opening of the Hartford & New Haven Railroad, which began providing freight service between New Haven, Hartford, and Springfield in 1844, would eventually eclipse the canal's usefulness for transporting freight. The arrival of the railroad led to a long-term decline in shipping upon the canal. The cessation of the canal's transportation functions was not, as suggested by some historians, instantaneous. This article appeared in *The Hartford Courant* on October 6, 1845, toward the end of the first season of canal operations after the opening of the railroad:

Business on the River

It was proposed by many, that the opening of the Rail Road to Springfield and the very low price of freight offered, would divert a greater portion of the business from the Boats up the River and transfer it to the Cars. A large amount of transportation has gone by the Rail Road, but the business on the River has increased notwithstanding.

The average amount of tonnage passing up the River in Boats – as estimated by the Collector at Windsor Locks, for the past nine years, is 16,225 tons annually, which does not include any of the freight landed on the banks of the Canal.

The tonnage of the present year is as follows, viz: passing up through the Canal at Enfield Falls showing an average increase of business over any period during the last ten

In April,	1256 tons.
" May	1753 "
" June	1936 "
" July	1416 "
" August	1863 "
" Sep.	2130 "

10,354 tons in 6 months.

---

[24] Charless Dickens, *American Notes for General Circulation and Pictures from Italy*, (London, Chapman & Hall, 1913).

[25] Connecticut River Company Records, Connecticut Historical Society.

[26] "Navigation of the Connecticut River," *The Hartford Courant*, August 14, 1868.

years. Add to this 1130 tons landed on the banks of the Canal, and the aggregate amount of merchandize transported up the River in Boats during the last 6 months is 11,434 tons. No definite estimate can be formed of the amount brought down the River, but it is probably larger than the up freight.

If the works for improving the Navigation on the River are kept in good condition, there is no doubt but the boating business will increase the more rapidly, the further up the valley the line of Rail Roads are extended, and the business of our City will increase in proportion. The Rail Road will open facilities for public travel, and for the rapid and certain transportation of the lighter and more valuable kinds of merchandize, and much of the heavy freight will go up on the river during the season of open navigation.[27]

The prediction that the canal would continue to serve as the preferred mode of transportation for heavy freight appears to have been accurate, at least to some degree. In 1886, some 42 years after the arrival of the railroad, Jabez Haskell Hayden wrote:

There are three or four freight-boats and a steam-tug plying between Hartford and Holyoke, and about the same number of large scows, which bring coal and some other heavy freight to this place.[28]

Haskell also reported in *Historical Sketches* that he received coal for his silk mill via the canal as late as 1885, when the "short haul bill" was passed.[29]

Certain products, such as gunpowder produced at the Hazard Power Company in Hazardville, were considered too dangerous to be transported by rail and continued to be transported via the canal for many years. That includes all powder sold to the Union army during the Civil War.[30] The members of the Windsor Locks Board of Selectmen were so concerned about the dangers to the community posed by the transportation of gunpowder from the Hazard Powder Mill through the canal that they passed an ordinance regulating the practice on March 16, 1875, some 31 years after the advent of the railroad.[31]

---

[27] "Business on the River," *The Connecticut Courant, October 6, 1845.*

[28] Hayden, Jabez Haskell, *Memorial History of Hartford County 1633 – 1884, Volume II*, Trumbell, J. Hammond, Editor, (Edward L. Osgood, Boston 1886) p. 568.

[29] Hayden, *Historical Sketches*, 98.

[30] Christopher J. Kervick, "Transportation of Gunpowder on the Windsor Locks Canal," Windsor Historical Society, 2012, *https://windsorhistoricalsociety.org/a-booming-business-transportation-of-gunpowder-on-the-windsor-locks-canal/*.

[31] Ordinance, Windsor Locks Board of Selectman Vol. 6, P. 88 of the Windsor Locks Land Records.

Logbooks within the Connecticut River Company Lock Tender records from the turn of 20th century, record regular passages of boats. Many of these passages tended to be seasonal, as upriver pleasure boat owners piloted sailboats and wooden motor yachts to coastal marinas on Long Island Sound at the beginning and end of each boating season.[32] The following is a typical entry in the canal logbook for that period.

> **Saturday August 9, 1902.**
> Temp. 53 at 5:30 am. Wea. Clear WSW
> Good Canal all night & day.
> Water in river dropped 2 ½ from 6 pm to 6 am.
> From 6 am to 6 pm at 2 and still.
> Total fall on this rise 5 in at head.
> 35 Turns of sucker at 7:30 pm
> Reading on gauge No. 1 38'2 ½" at 7:30 pm
> Passed - Sail Boat – No name down through locks.
> Passed – Steam Boat No. 3 Naval Reserve up through locks.
> Passed – Sail Boat No name up through locks.
>
> | Reading on Guage No | | | | No. of turns on Sucker | |
> |---|---|---|---|---|---|
> | 1 | 9:30 AM | 38 ¼ | | | 27 |
> | " | 2 | 9:33 | 38 ¼ | " | 27 |
> | " | 3 | 9:39 | 38 ¼" | " | 27 |
> | " | | 1 | No Readings Locking Boat | | |
> | " | | 2 | " | | |
> | " | | 3 | " | | |

Not only did the tender record the boats passed, but he regularly checked the water levels and adjusted the gates as needed to maintain a constant water level in the canal needed to efficiently supply water to power the mills.[33]

Boats were still, but rarely, passing through the canal into the mid-1970s.[34]

---

[32] C.H. Dexter Company Records, Series II, Subseries D, Connecticut River Company, Archives and Special Collections, University of Connecticut Library.

[33] Ibid.

[34] Author's own recollection.

CHAPTER ELEVEN

Industries Along the Canal

The Directors of the Connecticut River Company recognized early in the planning stages that the benefits of the Canal at Enfield Falls would not be limited to aiding navigation on the river. The water level at the southern end of the completed canal is thirty feet above the average water level of the Connecticut River. That difference represented potential energy. The Directors hoped to supplement the canal's toll revenue by leasing water rights to manufacturers who would be enticed to locate their mills on the site with the promise of a steady and predictable water supply.

The canal was designed to facilitate this dual purpose. In addition to the single guard lock to admit vessels into and out of the canal on the northern end, the design included a massive breast wall adjacent to the guard lock. This wall was equipped with sluice gates that enabled the canal's water level to be controlled with precision. The lock tender made hourly readings of the water level and cranked the gates up or down as needed to maintain a consistent water level. Industrialists interested in operating mills on the south end of the canal entered long-term leases for "water lots" on which to locate their mill and draw water from the canal. One or more apertures below the water line constructed in the side wall of the canal delivered water to each water lot. The water passed through the apertures into sluiceways, which directed the water over mill wheels before it was discharged into the river. The amount of rent paid for the water was determined by the size of the aperture.

Pre-Canal Mills

When the canal was constructed between 1827 and 1829, there were no existing local mills deriving power from the Connecticut River. Dexter's saw and grist mills drew their motive power from a Mill Pond, which was formed by damming the mouth of Kettle Brook and supplemented by the redirection of Add's Brook so that it discharged into the Mill Pond. Haskell's gin distillery drew its power from a small pond, also fed by Add's Brook. The canal proponents were confident that by providing the infrastructure - both transportation and power production facilities - for a large-scale manufacturing center, it was only a matter of time before enterprising industrialists would set up shop on the canal bank. In that sense, the area that would become known as Windsor Locks was Connecticut's first planned industrial park.

The Dexter and Haskell Mills

In 1769 Seth Dexter and Ephriam Haskell purchased 160 acres of land in the Pine Meadow section of Windsor for their sons, Seth Dexter *(locally, this Seth Dexter is often referred to as Seth Dexter I because he was the first to settle in Pine Meadow)* and Jabez Haskell. The families were connected by the marriage of Seth Dexter I to Jabez Haskell's sister, Deborah Haskell. The land was roughly bound to the north by what is now Grove Street, east by what is now Center Street, south by what is now School Street and west by the river. The land was transversed west to east by Kettle Brook and Adds Brook. The deed to Dexter and Haskell makes reference to an existing saw mill on the site. This mill was near the mouth of Kettle Brook and derived its power from a mill pond, located just west of the mill, which was formed by the damming of Kettle Brook and the redirection of Adds Brook into the mill pond in 1742.

Seth Dexter I and Jabez Haskell moved to Pine Meadow to live on this property in 1770. Immediately upon their arrival, the Dexters and the Haskells began to operate the saw mill on the mill pond. Seth Dexter I was a clothier and soon acquired the water rights to a mill and shop on Kettle Brook just north of its intersection with Center Street. Here, he began a cloth-dressing operation and began selling cloth from his shop. The clothier works were operated by the Dexters until sold to Timothy Mather in 1817. In 1784, the families added a grist mill to the enterprise, which was built next to the saw mill. The sons of Seth Dexter I and Jabez Haskell were Seth Dexter II, Harris Haskell, and Herlehigh Haskell. These men succeeded their fathers in the operation of the saw and grist mills and also expanded the family enterprises. Harris Haskell and Herlehigh Haskell began operating a gin distillery on the riverbank beginning about 1811. The Haskell gin mill was on the river across from the present entrance to Grove Cemetery. This mill was powered by water from a small pond, just to the west of the mill. The pond was supplied with water by a small diversion from Add's Brook, before it emptied into the main Mill Pond. Seth Dexter II, built a second grist mill, also on the riverbank, in 1819. The water to provide power for the second grist mill also came from the mill pond. When the canal was built, the raceway from the mill pond dam to the second grist mill had to be encased in a culvert running under the canal.

Charles H. Dexter, the son of Seth Dexter II, was born in 1810. As a young man, C. H. Dexter and William English began to experiment with different techniques for making paper in the basement of the Dexter grist mill using waste power. English had been an expert in the art of making paper by hand in Dublin, but his skills were no longer needed once the first paper making machines were introduced in Ireland. He came to the Windsor Locks in 1833 and began his friendship with C. H. Dexter. Each of the men became paper-making entrepreneurs.

English first operated a hand-made paper-making shop in an old mill leased from Timothy Mather at "mud dam" on Kettle Brook. This mill was the former clothier works operated by the Dexters. He then purchased a used paper machine and in 1845 moved his operations to a former saw and grist mill on Pine Meadow Brook *(now known as Waterworks Brook)*. This mill became known as the Hibernian Mill and was in operation until 1878, a decade following William English's death. It was operated by his son, James English, until its closure.

While conducting his experiments with English in the basement of the Dexter grist mill, C. H. Dexter developed the technique of making wrapping paper from manila hemp rope. This technique produced paper of exceptional strength. By 1840, he was producing 200 pounds of paper per day in the basement of the grist mill. Dexter is said to have obtained the fiber to make his "manila paper" from old jute bags used to deliver saltpeter to the Hazardville Powder Mills. To answer his need for more space and power, C. H. Dexter moved his paper-making operation to a wood frame building on the riverbank in 1840. There is no record of C. H. Dexter leasing water-power from the canal company at that time, so he most likely obtained power from the mill dam, via the culvert under the canal, which was already powering the grist mill.

Post-Canal Mills

Compared to transportation, the provision of water-power proved to be a slower developing but much longer lasting producer of revenue for the Connecticut River Company. The path of the proposed canal passed over the small pond fed by Add's Brook from which the Haskell gin mill gained its power, making the continuation of the pond impossible. In exchange, the Connecticut River Company offered to install a gate in the east wall of the canal in line with its bottom to pass water through a flume running under the tow path to the gin mill's water wheel. The Haskell Brothers readily accepted this offer as it provided a more reliable source of water than the pond had been providing. The mill could run around the clock with no time needed to recharge the pond. This arrangement proved to be the first instance of water from the canal used to power an adjacent mill, although the company received no income for the water, as it was given in exchange for the right to close up the pond.[1]

In 1833, Samuel Williams and Whiting Hollister opened a paper mill on the first water lot north of where the ferry road *(now Route 140)* crossed the canal.[2] They obtained a 999-year lease to draw water from the canal to provide power to their mill.[3] Dudley Persse and Horace Brooks acquired the mill soon after it began operation and added a second water lease in 1844.[4] For many years, one of the larger contracts of this com-

---

[1] Hayden, *Historical Sketches*, p. 33.

[2] Raber & Malone, *Historical Documentation*, p.70.

[3] 999 year Lease dated January 1, 1834, recorded February 5, 1834 in Vol. 33, Page 274 of the Windsor, Connecticut Land Records.

[4] 999 Year Lease dated March 6, 1844, recorded March 7, 1844 in Vol. 37, p. 50 of the Windsor, Connecticut Land Records; Raber & Malone, Historical Documentation, p. 70.

pany was for producing the paper used to print *The New York Herald.*[5] On August 30, 1845, *The Herald* printed a woodcut image of the Persse & Brooks mill on the canal and described their supplier's operations as follows:

PERSSE AND BROOKS PAPER MANUFACTORY.

As an appropriate accompaniment to the general description of the New York Herald establishment, we give an elegant and accurate engraving, representing one of the most extensive establishments of the kind in this country – the paper mills of Messrs. Persse and Brooks, 65 and 67 Nassau Street, New York. These mills are situated on the bank of the Connecticut river, at Windsor Locks, Conn. The adjacent country, forming a part of the celebrated valley of the Connecticut, is richly cultivated and presents a beautiful appearance in this season of the year. The main building is 135 feet long by forty wide, and two stories high with this to be reckoned an additional wing, 70 feet long by 40 wide, making the total length of the building 205 feet. The later building or wing is the machine and finishing rooms. The power which sets in work the machinery of this mill is a water power, and is derived from the Enfield Canal, which runs past the building, and as well as the railroad quite close to it, both giving great facilities for the transport of freight to and from the manufactory. The water power is one of the most valuable, retaining throughout the year a uniform and unimpaired force, more than sufficient for the end required, and is not affected perceptibly, even by the present great drought. The machinery set in motion by the water power, consists of two large water wheels, nine engines two 62 inch Fourdrinier machines, and all the requisite appointments to make such an establishment thoroughly efficient. There are from 30 to 40 hands constantly employed in various manual processes. The mill is capable of turning out 20,000 pounds of paper per week on the average throughout the year, and the annual value of the article manufactured amounts to $125,090 at least. The consumption of raw materials in the shape of rags is from twelve to fifteen tons a week: three-fourths of these are of home production. These otherwise worthless commodities are transformed with amazing celerity at this mill into paper, an article essential to civilized life, and subserving more than any other the most important purposes of man. A couple of days is the usual time taken to metamorphose some tons of rags into paper ready for use; so that the reader of this may hold in his hand what but a few days before was a tattered garment. But although this is the usual time taken in the manufacture of paper, the foreman informed us that in four hours he can turn out paper ready for use, but necessarily inferior in quality to that which is subjected to the bleaching process for a longer time. In the manufacture of paper, the rags are first sorted in the rag room, then dusted; they are next

---

5 "Persse & Brooks Paper Works," *New York Herald*, August 30, 1845.

conveyed to an implement called a rag-cutter, and cut up in small pieces to prepare them for the action of the boiler; in the latter they are boiled for twelve hours, and afterwards washed and purified of all useless matter; the steep chest next receives them, in order to be bleached; from thence they go into the beating engines, where they are belabored until reduced to a pulp, of proper consistence and quality. After all this, the material is ready for the machine, to be transformed into paper, cut, and finished. The distance to these mills, from New York, is 136 miles which is travelled by steamboat and railroad, in seven hours. They are halfway between Hartford and Springfield, and twelve miles distant from each of those towns.

This establishment furnishes us with all our paper. Since we have begun to receive supplies from it, we have paid to Messrs. P. & B. over $300,000, and we now pay them from $700 to $1000 a week.[6]

In 1834, Ebenezer Carlton leased a water lot from the Connecticut River Company to open a sawmill on the canal bank, using water-power from the canal.[7] This site was the first water lot south of the waste weir and the northernmost of all water lots. Ebenezer Carlton was a lumberman from Bath, New Hampshire who had supplied much of the lumber used in the construction of the canal. The site was convenient for Carleton, as it was directly across from the lower basin, where logs cut from his forest land around Bath, which had come down the river in rafts, could be stored until they were cut into dimensional lumber in the mill. The lower basin became known as Carleton's basin.

In 1857, Persse and Brooks opened a considerably larger facility on the water lot that had previously hosted Ebenezer Carlton's sawmill. The company assumed Carlton's lease. For a time, Persse & Brooks operated both the lower mill and the new upper mill simultaneously.[8] By 1864, Persse & Brooks had been taken over by the Seymour Paper Company.[9] In 1882, the Seymour Paper Company consolidated its operations in the large mill by the lower basin and expanded its water rights on that site.[10] The original lower mill was for a time used as a wool scouring mill

---

[6] "Persse & Brooks Paper Works," *New York Herald*, August 30, 1845.

[7] Raber & Malone, *Historical Documentation*, 70.

[8] Ibid.

[9] Hayden, *Memorial History of Hartford County 1633 – 1884,* Vol. II, Trumbell, J. Hammond, Editor, (Boston, Edward L. Osgood, 1886), p. 570.

[10] 999 year lease dated March 20, 1880 and Recorded April 8, 1880 in Vol. 6, Page 301 of the Windsor

by Dwight Skinner but in 1894 returned to the production of paper as the Anchor Mills Paper Company, which made several varieties of copying tissue paper.[11] The Anchor Mill was destroyed by fire on February 13, 1916.[12]

On the second water lot north of the canal bridge, Jonathan Danforth established a door butt factory in 1831. The lot was then subdivided and several mills operated there over the next fifty years, including Slate & Brown's machine shop, A. C. West's sewing machine company, the Denslow & Chase rifle company, and J.P. and H. A. Converse's original foundry.[13] Each of these mills used water power from the canal to some degree, but no individual leases to these operators were ever recorded in the land records. The site was then consolidated and used as a wool grading and washing works by A. Dunham & Son, later by H. R. Coffin, and lastly by Dwight Skinner, until he acquired and moved his operation to the first water lot.[14]

On the third lot north of the bridge was the Eli Horton & Son Company. Eli Horton was a skilled machinist who in 1855 obtained a patent for his "universal lathe chuck." A lathe chuck is the device that holds the workpiece in place as it is turned on a lathe machine. Prior to the development of the universal chuck, the machinist had to individually tighten several jaws around the perimeter of the chuck to secure the workpiece. Horton's universal chuck permitted the tightening of all jaws at the same time. This invention saved the lathe operator a great deal of time. Horton commenced business in 1856, possibly in space leased from Persse & Brooks.[15] In 1865 he constructed his attractive three-story mill and obtained a 999 year lease for water rights from the canal.[16] By this time, the traditional water wheel had given way to the more efficient

---

Locks, Connecticut Land Records.

[11] *The Hartford Courant*, November 16, 1893.

[12] *The Windsor Locks Journal*, February 18, 1918.

[13] Raber & Malone, *Historical Documentation*, p. 70.

[14] Ibid.

[15] Ibid.

[16] 999 year lease dated April 1, 1865 and Recorded March 14, 1873 in Vol. 6, p.12 of the Windsor Locks, Connecticut Land Records.

water turbine. Horton's original turbine produced 18 horsepower.[17]

On the fourth water lot north of the bridge, Joseph P. Converse and Hannibal A. Converse constructed a new iron foundry in 1844.[18] The foundry was later operated by A. W. Converse before the building and water rights were acquired by the Eli Horton & Son Company in 1873, the same year that Eli Horton & Son Company went public. With this acquisition, Horton added an iron foundry to its lathe chuck factory, each with its own water rights.

During this same period of time, several manufacturing establishments had sprung up south of the canal bridge. Among those were the Medlicott Company in 1864, a manufacturer of undershirts and drawers; C. H. Dexter & Sons, whose paper producing operation was established in 1836 and whose mill on the canal bank was constructed in 1840; the Farist Steel Company established in 1865; and Jabez Haskell Hayden's company for the production of silk thread from silkworms built in 1838 on the site of the Haskell gin distillery.[19]

At the southern terminus of the canal, situated just above the three lift locks, the Connecticut River Company built a factory on speculation in 1846. The mill was leased to a cotton thread maker known as Connecticut River Mills in 1847.[20] The main building was four stories tall and measured eighty feet wide by forty feet deep. Next to the main building was the two-story picker house. It was a large mill complex for its time and employed fifty people. The mill produced more than 2500 pounds of cotton thread per week.[21]

In 1871, J. R. Montgomery came to Windsor Locks and immediately leased the Connecticut River Mills plant.[22] There, he continued to produce cotton warp thread, and the company became a large re-

---

[17] Windsor Locks Bi-Centennial Commission, *The Story of Windsor Locks,* 1976; Raber and Malone, *Historical Documentation,* p. 80.

[18] 999 year lease dated December 27, 1844 and Recorded April 15, 1845 in Vol. 37 Page 85 of the Windsor, Connecticut Land Records.

[19] Raber & Malone, *Historical Documentation,* p. 70.

[20] Raber & Malone, *Historical Documentation,* p. 70.

[21] *Webb's New England Railway and Manufacturers' Gazetteer,* (Providence Press, 1869).

[22] Hayden, *Memorial History of Hartford County.1633 – 1886,* Vol. II, Trumbell, J. Hammond, Editor (Edward L. Osgood, Boston 1886), p. 570.

gional supplier. J. R.'s brother, George Montgomery, joined the firm in 1885. The duo proved to be an ideal mix of technical skill and creativity. George quickly expanded the firm's products to novelty yarn. These were yarns with metallic or other colorful and bright highlights twisted in and yarns with unique weave patterns. They were decorative and frequently used in trimming clothing. As soon as some new fashion or style came in from Europe, the Montgomery brothers would build a new machine or retrofit an existing machine to recreate the look, and then mass-produce it. Over time, the Montgomery Company was able to offer more than a thousand styles of novelty yarn.[23]

The rapid growth of the business resulted in the need for more manufacturing space. When Dwight Skinner moved his wool scouring operation from the second water lot north of the bridge to the first water lot, the vacancy offered the opportunity for expansion the Montgomery brothers sought. The Montgomery Company purchased the second lot and its water rights, removed the existing buildings and erected a four-story brick factory in 1881. That factory *(Building One)* was soon enlarged with the addition of a fifth story, and the addition of a second five-story brick building abutting Building One on the south side *(Building Two)*. All brick portions of the current building *(Buildings One and Two)* were in place by 1904. The J. R. Montgomery Company was now operating large mills at two locations on the canal - the original cotton thread mill by the locks, and the two newer brick facilities on the second water lot north of the bridge.[24]

At the of the turn of the century, there were four major industries on the canal bank north of the bridge. The Anchor Mills Paper Company was on the first lot. The J. R. Montgomery Company occupied the second lot. The Eli Horton Company operated its lathe chuck mill and foundry on the third and fourth lots, and the Seymour Paper Company was on the water lot farthest north. Although steam and electric power were driving most, if not all, of their machines by that time, each company retained its rights to draw water from the canal.

Both the Eli Horton & Son Company and the Montgomery Company proved to be enduring enterprises. Eli Horton died in 1878, but the

---

23 *J. R. Montgomery Product Sample Catalog,* Circa 1920, in Possession of Town of Windsor Locks.

24 J.R. Montgomery Company (1905), "Historic Buildings of Connecticut," https://historicbuildingsct.com/j-r-montgomery-company-1905/

firm was carried on principally by Ezra Brewster Bailey, who had married Horton's daughter Katie in 1871. Under E. B. Bailey, the company continued to refine its sole innovative product, the universal lathe chuck. It was offered in a variety of sizes that allowed it to hold workpieces as small as a drill bit and as large as a railroad car wheel. The market for its products was worldwide. E. B. Bailey was succeeded by Eli Horton's grandson, Ellsworth Horton, in 1912. In 1949, the Company merged with Gabb Manufacturing Company, a producer of aircraft parts. The plant was mod-

Postcard of buildings 1 and 2 of the J.R. Montgomery Company, which produced cotton wrap thread and mercerized yarn.

ernized to enable it to compete in the post-war era. Ultimately, the company became known as Connecticut International, with its primary product being airport runway lights. Connecticut International shut down its Windsor Locks operations in 1981. The properties were purchased by its long-standing neighbor, the Montgomery Company.

While the Eli Horton & Son Company prospered on the strength of one innovative product, the J. R. Montgomery Company succeeded by adapting its products to a wide range of demands and uses. Its ability to quickly adapt its machines to manufacture products that responded to a new market demand was the key to its success and longevity. In 1896, the Company introduced "mercerized" cotton which dramatically increased the demand for its cotton thread products. Mercerizing was a process which resulted in brighter and more flexible thread. In the early part of the twentieth century, the Montgomery Company began to produce tinsel thread, which was made by reducing various metals to foil thickness, and then winding narrow strips of the tinsel around cotton thread. At first, the tinsel thread was used in decorative dress or drapery fabrics. Eventually, as the conductive properties of the thread were discovered and refined, the thread was used in such products as telephone handset cords and Christmas decorations.

In 1920, the Montgomery Company purchased the remains of the Anchor Mills Paper Company building, its neighbor to the south. It razed the existing structure and replaced it with the white reinforced concrete building which remains to this day *(Building Three)*. The Company consolidated all of its operations into the two abutting brick factories and the new concrete factory. The old Connecticut River Mills buildings by the locks were abandoned and torn down shortly thereafter. With the purchase of the Eli Horton buildings and water lots in 1981, the Montgomery Industrial Complex reached its greatest size. The Company continued in operation until 1989. The next year, the Montgomery Industrial Complex was added to the National Register of Historic Places. The original Eli Horton & Son Company mill was the jewel of the buildings within the complex, both for its architectural beauty and its historic significance as the oldest surviving mill on the canal bank. Sadly, the building burned to the ground in a dramatic fire in 2006. Buildings One, Two, and Three have been converted into an attractive residential apartment complex with commanding views of both the canal and the river.[25]

On the first water lot south of the bridge, brothers John F. and James H. Wells operated a paper mill commencing in 1838.[26] After the mill was transferred to C.P. Hollister in 1843, it burned in 1847.[27] The mill was rebuilt in 1849 and operated by brothers Anson Blake and Gardner R. Blake under a new lease from the Connecticut River Company.[28] The Blake brothers produced cotton batting at the mill until it was sold to Lucius B. Chapman and operated as a stockinet factory. The Medlicott Company, a manufacturer of undershirts and drawers, began operations

---

[25] Montgomery Mill Apartments: Modern Meets Industrial, Hartford Courant, August 8, 2019; Montgomery Mills, Windsor Locks, https://crosskey.com/projects/historic-preservation/montgomery-mill/

[26] 999 year lease, April 10, 1838 and Recorded September 14, 1839 in Vol. 35, p. 278 of the Windsor, Connecticut Land Records.

[27] Raber & Malone, *Historical Documentation,* p. 70.

[28] 999 year lease dated April 28, 1849 and recorded May 19, 1840 in Vol. 37, Page 148 of the Windsor, Connecticut Land Records.

in the mill beginning in 1864. The Company was founded by William G. Medlicott of Longmeadow, Massachusetts, in 1846. It was originally located at Thompsonville, Connecticut. In order to take advantage of the ample water power available at the Enfield Falls Canal, Medlicott moved all operations to that site in 1864.[29] He assumed the 999-year lease for the water lot that had been entered into by the Blake brothers in 1849.

In addition to his business acumen, William Medlicott was an avid collector of rare books.[30] As a consequence of increasing business debt, he was forced to sell much of his collection in 1876. A catalog of his collection prepared to aid in that sale listed nearly 4200 titles and 6950 volumes.[31] The sale of the collection enabled him to keep the Medlicott Company out of bankruptcy, but the workout plan required that he liquidate his interests in the Company. His shares were sold to Charles C. Chaffee.[32] Under Chaffee's guidance, the business continued to flourish, and Medlicott wool underwear *(as uncomfortable as that may sound to modern sensitivities)* became an industry standard during the last quarter of the 19th century and first quarter of the 20th century. Eventually, as the industry evolved and more comfortable materials were demanded by consumers, the Medlicott Company was liquidated during the 1940s.[33] From 1950 to 1952 the mill was owned by the LaPoint Plascomold Corporation and used for the manufacture of antennas and other television accessories.[34] In 1963, the property was purchased by its neighbor to the south, C. H. Dexter & Sons, Inc. and ultimately torn down to make room for the expansion of the Dexter mill and warehouse facilities.[35]

In partnership with his brother-in-law Edwin Douglass, who was introduced to his sister while serving as the Assistant Engineer during the construction of the Enfield Falls Canal, Charles H. Dexter established

---

[29] Obituary of William G. Medlicott, *Springfield Republican*, February 20, 1883.

[30] R. J. Hall, "William G. Medlicott (1816-1883): An American book collector and his collection." (*Harvard Library Bulletin*, Spring 1990).

[31] Ibid.

[32] *The Morning Journal-Courier* (New Haven, Connecticut), October 31, 1900.

[33] *The Hartford Courant*, September 8, 1950.

[34] *The Hartford Courant*, July 2, 1952.

[35] *The Hartford Courant*, December 21, 1963.

12356 — Medlicott Company, Windsor Locks, Conn.

C. H. Dexter and Company in 1847.[36] The Company expanded its paper mill capacity at that time, and therefore increased its water-power requirements. The Company met that need by entering into a 999-year lease with the Connecticut River Company.[37] The original mill on the canal was destroyed by fire in 1873 and a new brick mill was completed and began operations in 1875. This new mill is now the core building in the large paper-making complex that remains on the canal to this date. The company now owns all the land and water-lots from the Route 140 bridge south to the locks. Following a remarkably successful history of product innovation, physical plant expansion, and workforce growth, the Dexter Corporation, as it was then known, was sold to Ahlstrom Paper Group, a Finnish paper-making company, in September, 2000. The mill remains in operation on the canal bank.

The gin mill of Harris and Herlehigh Haskell that began operations on the riverbank in 1811 was still operating when the canal was built. Several of the mill outbuilding were moved and their mill pond filled in to make way for the canal. The road leading to the gin mill also had to be moved to the west of the canal. The Connecticut River

---

[36] Raber & Malone, Historical Documentation, 70.

[37] 999 year lease dated March 8, 1840 and Recorded January 13, 1911 in Vol. 16, p. 132 of the Windsor Locks, Connecticut, Land Records. It is unclear as to why this lease was not recorded in the land records until nearly 61 years after its execution.

Company was forced to build a bridge from the relocated road across the canal to provide continued access to the Haskell gin distillery. The gin distillery was also allowed to draw water at no cost from the canal using a "deep conductor" that was installed in the wall of the canal in line with its bottom. The distillery was converted to a silk thread mill in 1838. A lease to draw water from the canal was secured by the Haskell brothers in 1842.[38] The lease differed from all other water-lot leases in that it had no specified term. Perhaps that is because the Haskell's already owned in fee the land on which the mill was located. The lease also contained the agreement that the Haskells would not use the "deep conductor" longer than an additional five years. The silk mill operated on the site until 1895.[39]

A new post office to serve the needs of the growing industrial area was opened in 1843. The U. S. Postal service appointed C.H. Dexter as the new postmaster. The industrial area of Windsor around the southern end of the canal had become known as "the Locks." At Dexter's suggestion, the name Windsor Locks was used to identify the new post office. The village desired to be separated from Windsor and was incorporated as the Town of Windsor Locks in 1854. During this period, the name Enfield Falls Canal fell into disuse and the name Windsor Locks Canal became popular. C. H. Dexter was named President of the Connecticut River Company in 1855, and he greatly accelerated the development of the water-power aspects of the business.

Conclusion

The merchants and industrialists who envisioned, promoted, and financed the canal have been rightfully lauded for their efforts. The canal was, after all, a reflection of the tremendous entrepreneurial spirit of these men. There is limited recorded information about the canal in general, and the little that does exist, is disproportionately focused on those who envisioned and financed the canal. There is scant mention of the bright and creative young engineers who designed the

---

[38] Lease dated March 16, 1842 and Recorded June 10, 1843 in Vol. 37, p. 4 of the Windsor, Connecticut, Land Records.

[39] Raber & Malone, *Historical Documentation*, p.70.

canal and oversaw its construction. There is even less information about the humble men who toiled *(and in some cases gave their lives)* building the canal. This imbalance is much less the result of overt prejudice or callous indifference, as it is a reflection of the economic and social fabric of Connecticut at the time the canal was built.

The historical reality is that the Windsor Locks Canal, and the other ambitious public works of the era, could not have come into existence without an amalgamation of all these groups. Connecticut and America were forever changed as a result. The canal should therefore be viewed as a bellwether of the emerging American experience. This new America relied upon the best efforts of all, both its citizens and its newly arriving dreamers, each bringing to the table their God-given talents. Combined, they created an America that became bigger than the sum of it parts. That is the canal's legacy and a reminder of the respect owed to every person who had a hand in its creation.

APPENDIX

Irish Surnames
Associated with the Windsor Locks Canal

Barry
Bostwick
Brennan
Brunson
Burk, Burke
Butler
Cain, Kain
Cardle
Carroll
Casey, Casay
Collins
Connery
Coogan, Cogan
Cosler
Costello, Costlo
Coulter
Couron
Cragan, Cregan
Guinney, Guinnane, Guinang
Curry
Deland
Dillon
Donoho, Donahue
Doran
Doyle
English
Farrel, Farrell, Farrelly
Finley

FitzGerald
Fitzpatrick
Fox
Furly
Gates
Gilday
Gilligan
Glughan
Grinnel
Haley, Haly, Halley
Hayes, Hays
Henderson
Hennessey, Hannassey
Holden
Howard
Jenks
Kennan
Killbride
Lane
Lannigan
Lynch
Maguire, McGuire
Maloney
Manning
McCarthy
McKenna
McMahon, McMan
Moore

Moran, Moron, Morron	Rooney
Morris	Ross
Murphy	Ryan
Murray	Shannon
O'Cavanaugh	Sheedy
O'Brien	Sliney
O'Connel, O'Connell	Smith
O'Dee	Stewart
O'Neill, O'Neil	Sullivan
Outerson, Outersen, Auterson	Timmons
Pelfory	Towers
Persse	Travers
Power	Wallis, Wallace
Quinn	Walsh

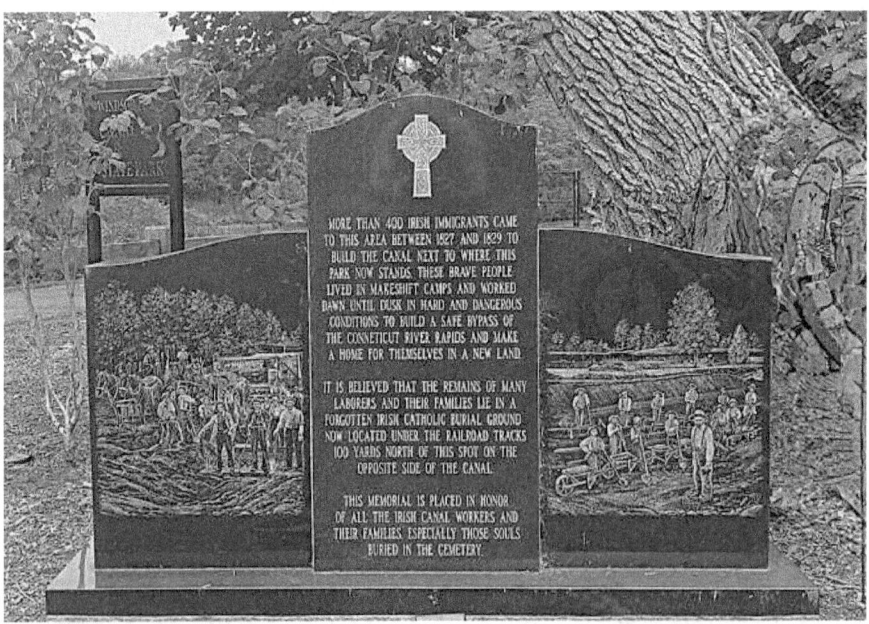

Erected by Town of Windsor Locks and the Ancient Order of Hibernians in 2022, this Irish workers' memorial commemorates the 400-plus Irish immigrants who labored to build the Windsor Locks Canal.

INDEX

A

Abbe, Charles, 155
Abbe, Jr., Charles Irish day laborer, 155
Abbe, Nelle Grace, assessment of Irish laborers, 119; Farmington Canal flood, 146; A. C. Western Sewing Machine Company, 177
Add's Brook, 11, 172 - 174
Adrian I, asserts supremacy over Ireland, 28
A. Dunham & Son 177
Agawam (steamer)
  and Charles Dickens, 164-166
Ahlstrom Paper Group, 183
Allen, James, Irish day laborer, 155
American Notes (book), Charles Dickens records trip aboard *Amagam s*teamer, 165
American Revolution
  impact on Irish 31-32
Anchor Mills Paper Company, 177
  purchased by Montgomery Company, 181
Archdiocese of Boston, 12
Association for Improving the Navigation of the Connecticut River 63
Atwater, Lyman, 110; biography 109, 110, 134, 143; canal contractor 109
  real estate speculator 109, 110, 134, 143
Atwater, Mehitable, wife of Eli Bronson, 110
Averill, Eliphalet, 61

B

Bailey, Ezra Brewster, lathe chuck manufacturer, 180-182
Baldwin, Laommi, 37; develops hydraulic cement, 39
Ballinvreena, Ireland, 33
Baltimore, MD, 15
Barker, Col. Mason, masonry contractor, 102
Barkhamsted, CT, 110
Barnet (steamboat) 72 - 79
Barnet, VT, 15, 54, 63
Barry, Jonathan, Irish day laborer, 127
Battle of Clontarf, 28
Beach, George, 66

Belcher, Samuel, 102
Bellows Falls Examiner, endores Connecticut River improvements in NH and VT, 69, 149
Bellows Falls, VT, 35, 37; and *Barnet*, 77
Bissell, Aaron, petitions for river improvement, 25
Bissell, Elisha, petitions for river improvements, 25
Bissell, John, 21
Bissell, Josiah, 43
Bissell, Orren, Connecticut River Company contractor, 157
Bissell's ferry, 22
Blackstone Canal, RI, 71
Blake, Anson, paper mill owner, 181
Blake, Gardner R., paper mill owner, 181
Blanchard, Thomas, 166; inventor, 148-149; steamship builder, 147-152, 148-149
Block, Adrian, 16
Boru, Brian, 28
Boston, MA, 15; canals, 86; difficulty transporting goods, 37; navigable rivers, 40
Brace, Thomas, 66
Brehon laws, 29 - 30
Brennan, Ellen, Irish immigrant, 128
Bridge inspectors, 43
Briggs, Greene, 155
Briggs, Munroe, Irish day laborer, 155
Bronson, Eli, father of Philo, 110
Bronson, Philo, biography, 110; contractor 109
Bronson, Zebrina, Connecticut River Company contractor, 157
Brooks, Horace, paper mill owner, 174 - 176
Buck, Daniel, 61, 66; Association for the Improving the Navigation of the Navigation River, 61, 66
Burke, Edmund, Connecticut River Company subcontractor, 155
Butler, Richard, Irish day laborer, 128
Cain *(Kane)*, Timothy, Irish day laborer, 155
Canal Leaks, 163 - 164
Canallers, 55, 74, 75; verbal battle with

Riverites, 85
Canal locks, 45
Canals, 35
Canal transportation, 164 - 170
Cape Cod Canal, 46
Captain Strong, co-captain of steam ship *Barnet*, 78
Cardle, Elizabeth, Irish immigrant, 128
Carlton, Ebenezer, sawmill owner, 176
Carroll, Thomas, railroad subcontractor, 121
Catholics 2 8- 29, form the Irish Defenders 31-32; low tolerance for, 124
Cayuga-Seneca Canal (NY), 110
Chaffee, Charles C., 182
Chamberlain, James, 22
Chapin, M. W., 78
Chapman, Lucius B., stockinet factory, 181
C.H. Dexter, 13, 182; history, 183; paper mill, 178
Chenango Canal (NY), 110
Cheverus, Rev. John, Bishop of Boston, 124 - 125
Clinton, DeWitt, Erie Canal, 35; reviews Farmington Canal, 59, 86
"Clinton's Ditch", 47
Cohoes Bridge Company, 103
Colebrook, CT, 54, 55
Confederation of Kilkenny, 30
Connecticut Canal Historic Exhibition Center, 12
Connecticut Department of Environmental Protection, 11; Bureau of Parks and Forests, 12
Connecticut General Assembly, 41- 42; approves petition for East Windsor river improvements, 25; authorizes widening Connecticut River, 18, 36, 38, 42, 54, 55, 63 ,66 - 69, 85, 86, 88, 91, 110, 113, 129; passes act to incorporate Farmington Canal, 54; regulates toll bridges, 41
Connecticut River, 12, 16, 17, 21, 54; bridges over, 23; channel widened, 15, 16, 17, 23, 36, 37, 40, 41, 43 - 45, 54, 61, 63 - 70, 71, 73 - 79, 82, 83, 85 - 88, 93 - 97, 100 - 102, 104, 105 - 108, 110 - 115, 18, 120, 126, 129, 139, 143, 145, 146, 148, 149, 150, 151, 152, 156, 162, 164, 165,167, 171,172, 173,175, 176, 181; survey of, 70; toll bridge over, 41
Connecticut River Banking Company, charter issued, 83
Connecticut River Company, 13,15,37,41 - 42, 45,66 - 68, 70, 71, 72, 73, 76, 78, 83, 84 - 99, op. cit. ,101 - 104, 105, 108 - 110, 112 - 118, o. cit., 120 - 122, 131 - 137, 138, 140, 141, 142 ,143, 147, 150, 153 - 155, 157, 158, 159, 161- 163, 166, 168, 171, 174, 176, 178, 181, 183, 184: purpose of 15; difficulty transporting goods, 37
Connecticut Steamboat Company, 62
Connery, Michael, Irish day laborer, 155
Converse Iron Foundry, 178
Converse, Joseph P., iron foundry, 178
Coolidge, Carlos, Asst.secretary, Vermont Convention, 69
Cornell University, 12; Enfield Falls Canal, 12
Cornish & Co., supplies labor to Connecticut River Company, 109
Cosler, William, Irish laborer, 128
Costello, Michael, 32,34; headstone, 130
Couron, Jonathan, Irish laborer, 128
Cromwell, Oliver, 30
Cumberland Canal, ME, 71
Curry, Hugh, Irish laborer, 128
Curry, Mary, wife of Hugh, 128
Curry, Patrick *(infant)*, 128

D

Daggett, David, Supreme Court justice, 91
Danforth, Jonathan, factory owner, 177
Davenport, Reverend John, 51
Davis, Ambrose, Connecticut River Company contractor, 157
Defenders, Irish revolutionary group, 31
Deland, Danie, Irish day laborer, 155
Delaware and Chesapeake Canal, 48
Delaware and Raritan Canal, 48, 72; and cholera, 122
Delaware Plantations, 52
Delaware River, 52; and Fulton, Robert, 62
Denslow & Chase Rifle Company, 177
Dexter, Charles H., paper mill owner, 11, 101, 173 - 174, 184; president of Connecticut River Company, 184
Dexter, Harriet C., marries Edwin Douglass, 101

Dexter, Seth, 172-173; grist mill damaged by canal leaks, 163
Dexter, Seth, II, mill owner, 173
Dexter's Saw and Grist Mills, 172 - 174
Dickens, Charles, on trip along Windsor Locks Canal, 165 - 167
Dillon, Charles, 121
Directors of the Association For Improving the Navigation of the Connecticut River, 66
Disease, and Irish laborers, 122
Donoho, John, Irish day laborer, 155
Doran, Bridget, wife of William, 127
Doran, Mary *(infant)*, 127
Doran, William Irish day laborer, 127
Douglas, Edwin A., assistant engineer, Connecticut River Company, 99, 100, 101; joins inlaws' papermaking business, 101; update on Enfield Falls Canal, 100 - 102, 104, 106, 107, 109, 182
Doyle, Ann *(infant)*, 127
Doyle, Austin, Irish day laborer, 127
Doyle, Ellen, wife of Austin, 127
Doyle, Martin, Connecticut River Company subcontractor, 155
Dublin, Ireland, 30

E

Early Public Improvement Projects, 18
East Hartford, CT, 23
East Windsor, CT, 22, 24; inhabitants petition for river improvements 24 - 25
Eaton, Samuel, 51
Eli Horton & Son Company, 178, 179 lathe chuck company, 178
Ellsworth, H. L. , 61
Ely, William, 66, 72
Enfield Bridge Company, 45, 46, 47, 63; and Enfield Falls, 88; loses Supreme Court ruling, 91; opposes Connecticut River Company, 90-91
Enfield Falls Canal, 35, 48, 99, 100, 101, 184; head gate, 13; Mason Barker on, 12, 13 ,35, 48, 98, 99, 100, 101, 103, 104, 105, 106, 110, 111, 112, 113, 119, 122, 123, 124, 130, 131,135, 143, 145 - 147, 153, 163, 166, 180, 182; shipping 166 - 169
Enfield Falls, CT, 16, 35 ,45, 46, 63, 78, 88;

and Edwin Douglass, 100 - 101; locks suggested for, 79; shipping to, 16
English, James, paper mill owner, 171
English, William, paper mill, 171
Enterprise (steamboat), 62
Erie Canal, NY, 12, 35, 47, 53, 99, 103; barge on, 12; Chief Engineer, Benjamin Wright, 48

F

Farist Steel Company, 176
Farmington Canal Company, 54, 55; threat to Hartford business, 53, 54, 55, 56, 57, 58, 59 ,61, 85, 86, 90, 145, 146
Farmington, CT, 55,59
Farmington River Canal, 56; chief engineer 55; lawsuits, 56; sabotage of, 56
Farmington River, CT, 54; bridge, 23
Rev. Fenwick, Benedict Joseph, Bishop of Boston, 125
Fitch, John, use of steam power, 62
FitzGerald, John *(infant)*, 128
FitzGerald, Mary, wife of Michael, 128
FitzGerald, Michael, Connecticut River Company subcontractor, 155
Fitzpatrick, Dennis *(infant)*, 128
Fitzpatrick, Mary, wife of Thomas, 128
Fitzpatrick, Thomas, Irish day laborer, 128
Fox, Ellen, wife of Patrick, 128
Fox, Mary *(infant)*, 128
Fox, Patrick, Irish day laborer, 128
French Revolution, impact on Irish Catholics, 31 - 32
Fulton, Robert, steamboat travel, 62

G

Gabb Manufacturing, 178
Garret, Patrick Irish day laborer, 128
Gates, Nathaniel Irish day laborer, 155
General Court at Hartford, authorizes highway, 19
Gilbert, Eunice wife of Elisha Punderson, 110
Gilday, Frank, Irish day laborer, 155
Glughan, Jonathan, Irish laborer, 128
Goodwin, James 61; Secretary of Association For Improving the Navigation of the Connecticut River 61, 66
Governor Dewitt Clinton, 47
Granger, Bethena Brownson, 108
Granger, Rufus, 97

Enfield Bridge Company, 88
Grasso, Gov. Ella T., 20
Greenfield Gazette, supports Connecticut River improvements, 68
Grinnel, Daniel, Irish day laborer, 128
Grinnel, Jerusha, wife of Daniel, 128
Grinnel, Jon *(infant)*, 128
Charlotte Griswold, on Irish labor camps, 120
Griswold Fish Place, location of Irish labor camp, 120
Griswold, Gaylord, Connecticut River Company contractor, 157
Griswold, Shubael, 43
Griswold, Williston, Connecticut River Company contractor, 157; land damaged by canal leaks, 163
Joseph Groumly, killed aboard the *Barnet*, 78
Grove Cemetery, 173
Guinane, Caroline, 138
Guinane, Eliza, 138
Guinane, James, 138
Guinane, John, 138
Guinane, Mary, 138
Guinane, Michael, 138
  Connecticut River Company subcontractor, 155
Guinane, Thomas, 138

H
Halley, Michael, Irish day laborer, 155
Haly,Margaret, wife of Thomas, 127
Haly, Margaret *(infant)*, 127
Haly, Thomas, Irish day laborer, 127
Hampshire and Hampden Canal Company, MA, 85, 86, 87
Hampton (steamer), 166
Hancock, Governor John, 38
Harris and Haskell Gin Mill, history, 170 - 171, 176, 181-182
Hartford, CT, 16, 62; fur trading post, 16
Haskell, Elishabeth, and Irish labor camp, 120
Haskell, Ephriam gin mill owner, 172 - 173
Haskell, Harris, 97; Connecticut River Company contractor, 157; grist mill damaged by canal leaks, 163
Haskell, Herlehigh, 97; grist mill damaged by canal leaks, 163; mill owner, 173
Haskell, Jabez, 172

Haskell's Silk Mill, 11
Hathaway, Charles, 97
Hayden, Jabez Haskell, on engineers of the Connecticut River Company, 74, 100, 106, 119, 130, 135, 146, 160, 167, 178; on steam boat *Barnet*, 74
Hayes, Daniel, Connecticut River Company subcontractor, 154
Hayes, James, Irish day laborer, 127
Hayes, John, Connecticut River Company subcontractor, 154
Hayes, Jonathan, Irish day laborer, 128
Hayes, Joseph, Connecticut River Company subcontractor, 154
Hayes, Margaret *(infant)*, 128
Hayes, Mary Irish immigrant, 128; wife of Jonathan, 128
Hazard Power Company, concerns over shipping gunpowder, 168
Hennessey, James, Irish day laborer, 127
Henry II, asserts papal supremacy over Ireland, 28
Henry VIII, ends Irish land ownership, 29
Hibernian Mill, paper mill, 173
Hilhouse, James, 90; champions Farmington Canal, 53, 57, 86 - 87; on Irish labor 58
Holden, Jonathan Irish day laborer, 128
Hollister, Whiting, paper mill owner, 174-175
Hopkins, Edward, 51
Horton, Ellsworth, 180
Hosmer, Judge Titus, hears suit against Connecticut River Company, 89 - 91
Hosmer, Titus, 89
Hotchkiss, Clarissa, wife of Lyman Atwater, 110
Howard, Thomas, Irish day laborer, 128
H. R. Coffin Wool Grading Co., 177
Huit, Reverend Ephraim, Farmington River Bridge, 23
Hurd, Davis, chief engineer, Farmington Canal, 57
Hurd, Jarvis, chief engineer, Farmington Canal, 57
Hutchinson, Holmes, 71
  surveyed a portion of Connecticut River, 70
hydraulic cement, 49, 72

INDEX

I
Imlay, William, 61
industrial revolution, 16
internal improvements, 27
Ireland
 Norman-English invasion of, 28 - 29
 Penal Laws, 30
Irish, and Farmington Canal, 58; French Revolution impact, 31 - 32; built Hampden Canal 58; cheap labor force, 111
Irish laborers *(see Irish day laborers by Surname);* social isolation of, 123
Irish Parliament, enacts Penal Laws, 30-32
Irish potato famine, 27; evolution of Irish Poverty, 27
Ive's Bridge, 162
Ives, James, 155

J
Jencks, Charles, 22
Jenks, Richard, Irish day laborer, 128
John F. and James H. Wells Paper Mill, 181
J.P. and H. A. Converse Foundry, 177
J. R. Montgomery Company, 179, 180 - 182
 history of, 180 - 181

K
Kalter, Patrick, Connecticut River Company contractor, 155-156, 170
Keach, Hosea, 23
Kennan, James, Irish day laborer, 155
Kervick, Frederick, 11
Kettle Brook, 11, 172
Kibbe, Isaac, 40
Kilbourn, Henry, 66
Kilfinane, Ireland, 32, 33
King, Abel, Connecticut River Company contractor, 154 - 155
King Charles I, execution of, 30
King, Dan, 120, 162
King Henry II, 28
King Henry VIII, 29
King, Mary Ballantine, wife of Henry L. Loomis, 108
Kings Island, CT, 65, 95

L
Lake Memphremagog, VT, 63, 70
Lamberton, Elizur, biography, 160; bridge tender, 159
Lamberton, George, 52 - 53
Lanman, James, Supreme Court justice, 90
Lannigan, Catherine, Irish immigrant, 128
LaPoint Plascomold Corporation, buys Medlicott building, 180
Ledyard (steamer), 166
Lehigh Canal, PA, 48, 72, 99, 103
Lehigh Coal and Navigation Company, 100; names Edwin A. Douglass chief engineer, 100
Lester, Daniel, biography, 161; Connecticut River Company contractor, 157; lock tender, 160
Lock & Canal Companies, NH, 82
Lock Tender, duties, 157 - 158
Loomis, Alle, 97
Loomis, Amasa, petitios for river improvements, 25
Loomis, Henry, sued by Connecticut River Company, 108
Loomis, Nathaniel, 108
Lower Enfield Falls, CT, 13
Lubagh River, Ireland, 32
Lynch, Ellen, Irish immigrant, 127

M
Mad Tom Bar, 41, 44, 45
Maguire, Felix, Irish day laborer, 155
Malone, Patrick, 11
Maloney, John, Irish day laborer, 137, 155
Manning, Margaret, Irish immigrant, 127
Marten, Walter, 11
Massachusetts, charters Hampshire and Hampden Canal Company, 54
Mather, Oliver, 43
Mather, Talcott, Connecticut River Company contractor, 157
Mather, Timothy, 171
McMahon, John, 137, 138 - 139, Connecticut River Company subcontractor, 155
McMahon, Lawrence, 32,34; Connecticut River Company subcontractor, 121
Mechanics Bank of New Haven, 56
Medlicott Company, clothing manufacturer, 178
Medlicott, William, as book collector,182
Merrimack River, 40
Middlesex Canal, N, 38, 39, 46, 63

Miller's Falls, CT, 37
Millers Falls, VT, 35
Millier's Falls, CT, 64
Mill Pond, 11, 170, 171
Mill River, CT, 51
Montgomery, George, cotton mill manufacturer, 177, 178
Montgomery Industrial Complex, listed as National Register of Historic Places, 178 - 179
Montgomery, J. R., cotton manufacturer, 176 - 179
Montgomery Mill, 11, 178 - 179
Moore, Eli petitions for river improvements 25
Morey, Samuel, side paddle wheeler, 62
Moron, Enos, 97
Moron, Jerusha, 97
Moran *(Moron)*, John, Irish day laborer, 121, 186; land damaged by canal leaks, 163
Morris Canal, NJ, 36
Morris, Thomas, Connecticut River Company subcontractor, 155
Morron *(Moron, Moran)*, Enos, Connecticut River Company contractor, 157
Murphy, Daniel, Irish day laborer, 128
Murphy, Hugh, Irish day laborer, 155
Murphy, Michael, Connecticut River Company subcontractor, 155
Murray, John *(infant)*, 128
Murray, Mary, wife of Jonathan, 128
Murry, Jonathan, Irish day laborer, 128

N

New Hampshire, 17, issues operating permit to Connecticut River Company, 85; lack of reliable water route, 37
New Hartford, CT, 54
New Haven Colony, 51
New Haven, CT, rivalry with Hartford, 7, 44, 51- 62, 66,75, 86, 90, 110,126, 127,132-133, 145-147,166
New Jersey, New Haven Colony buys portions of, 51
New Orleans New Basin Canal, 122
New York City, NY, source of cheap immigrant labor, 111
New York Herald, paper made by Presse and Brooks , 173v-174
New York, NY, 15; and Farmington Canal, 87
Norris, Col. Samuel, 161; biography, 129
Norris, Nancy, wife of Col. Samuel, 129
Norris, Samuel, lock tender, 159
Northampton, MA, 54

O

O'Brien, Johanna, Irish immigrant, 128
O'Cavanaugh, Rev. Bernard, 129,1601 first Catholic priest in Hartford, 129
O'Connell, Daniel, Irish day laborer 155
O'Connel, Mary, Irish immigrant, 128
O'Dee, Levi, Irish day laborer, 128
Old Saybrook, CT, 62
Oliver, Charles Silver, 32; orders Patrick Wallis's execution, 33; United Irishmen target, 32
Oliver Ellsworth (steamship), 62
O'Neill, John, Irish day laborer, 155
O'Neill. Mary, Irish immigrant, 127
Onrust, 16
Orcott Falls, VT, 36 - 37
Oxford Canal, ME, 71
Oxford, NH, side paddle wheeler debuts, 62

P

Palmer, Captain, co-captain of *Barnet*, 78
Parker, Peter, bridge carpenter, 155
Parmelee, Darius, 97
Parsons, Orrin, day laborer, 155
Parsons, Quartus, day laborer, 155
Parmelee, Nathaniel, 97
Payson, Moses P., President of Vermont Convention, 69
Pease, Cyrus, day laborer, 155
Pease, Levi, day laborer ,155
Pease, Lois, married Marcellus Pinney, 160
Pelfory, James, Irish day laborer, 127
Penal Laws, 31-32; contribute to Irish poverty, 30-31
Persse and Brooks Paper Manufactory, 175-177
Persse, Dudley, paper mill owner, 174-175
Peters, John S., 98
Peters, John Thompson, Supreme Court justice, 90
Pettibone, Joseph, fishshed manager, 163; land damaged by canal leaks, 163
Phantom Ship (New Haven, CT), 53, 56

INDEX

Philadelphia, PA, 15
Pierce, Otis, Connecticut River Company contractor, 162
Pinney, Marcellus, biography, 160; lock tender, 159
Pinney, Oliver, 160
Pomeroy, James, lock tender, 159
Pomeroy, James Austin, biography, 160
Pomeroy & Lester, Connecticut River Company contractors, 161-162
Porter, David, 66, 72
Power, Rev. John, performs Catholic funeral, 130; visits Windsor Locks, CT Irish workers, 126
Pratt, Joseph, 61,78, 102
Proprietors of Locks and Canals on the Connecticut Rive, 36
Proprietors of the Enfield Locks and Channels, 46
Punderson, Elisha, biography, 110 canal contractor 109
Pynchon, William, 16

Q
Quechy Falls, CT, 37
Quinnipiac River, CT, 51

R
Ranelagh, John O'Beirne, *A Short History of Ireland*, 27
Raber, Michael, 11
Reading, MA, 38
Reynolds, John, 23, 40; and Enfield Falls 88; and toll bridges, 41, 44
Reynolds, Rev. Peter, 40
"Riverites", 54, 61, 66, 71, 74, 85
Rooney, Patrick, Connecticut River Company subcontractor, 155
Russ, John, 61
Ryan, Edward, Irish day laborer, 155
Ryan, Thomas, Connecticut River Company subcontractor, 155

S
Saw Mill Path, 11
Schuylkill Canal, 72
Seymour Paper Company, 174-175
Shannon, Catherine, Irish immigrant, 127
Sheedy, Roger, and Patrick Wallis, 33
Sherman, Zephidiah, Connecticut River Company contractor, 161
Sigourney, Charles, 61
Simonds, Tabitha, 120, 162
Sisitsky's Market, 11
Skinner, Dwight, paper mil owner, 177
   Wool Grading Company, 177
Slate & Brown's Machine Shop, 175
Sliney, Jonathan & Mary, Irish laborers, 127
Sliney, Mary, 127
Sullivan, John L., 63
Smith, Alfred, 63; Association for the Improving the Navigation of the Connecticut River, 9, 63, 64 ,66, 67, 68 ,70, 77,81 ,83, 98, 99,102, 107,108,112,113,116,118,153
Smith, Bridgit, wife of William, 128
Smith, Elihu, 161
Smith, Newton, biography, 160 - 161; Connecticut River Company contractor, 157; lock tender, 159
Social isolation, and Irish, 123
South Hadley Canal, 36, 37,63
South Hadley Falls, MA 35,37
Southington, CT, 59
Southwick, MA, 54, 59
Sperry, Elish, 97
Springfield, MA, 16, 54
State of Connecticut, Department of Environmental Protection, 11
Steam Power, on Farmington Canal, 62
Stewart, Abraham, Irish day laborer, 155
St. Finnian, 32
St. Munchin's Church, Limerick, Ireland, 34, 199
Stoneham, MA, 38
St. Patrick, 27
St. Peter's Church, New York, NY, 126
Suffield, CT, 13
Sullivan, James, canal supporter, 39; championed canal to Merrimack River Valley, 37,38
Sullivan, John L. 39,46; experiments with steam-propelled barges, 40
Sullivan, Mary, wife of Owen, 127
Sullivan, Owen, Irish day laborer, 127
Sullivan, Thomas *(infant)*, 127
Surf Bar, 45
Switch-Back Railroad (PA), 100
Syracuse and Oswego Railroad, 71

T

The Anchor Mills Paper Company, 179
The Connecticut River Banking Company, 13, 84
The *Fulton (steamboat)*, on Connecticut River, 62
The Great Meadow Drain, 18
The Greenfield Gazette endures river improvements, 68
"The Locks", 13
"The Phantom Ship", 53. See also New Haven.
Timmons, Patrick, Irish day laborer, 155
Tolls, Connecticut Legislature approves for cargo on the Connecticut River, 18, 20 - 24 op. cit., 40, 44, 60, 68, 88 - 89, 157
toll roads, 18, 20
Towers, William, Irish day laborer, 155
Turnpike Road *(Rte. 75)*, 20

U

Union Company, CT Assembly charters to improve Connecticut River channel, 25
United Irishmen, 31
Upper Enfield Falls, CT, 13
U.S. Supreme Court, rules against Enfield Bridge Company, 91
U.S. War Department, surveys Connecticut River, 70
Utica and Syracuse Railroad, 71

V

Vermont, 17; first to grant Connecticut River Company an operations permit, 84
Vermont Convention, 68; officers elected, 69; petitions U.S. to support Connecticut River survey, 70
Vermont (steamer), 148-149; trips on Enfield Falls Canal, 166

W

Wallis, Staker, see Michael Wallis, 33
Wallis, Patrick, 32; hanged and beheaded, 33; martyrdom of, 33
Walsh, Bridgit, Irish immigrant, 128
Walsh, Eleanor *(infant)*, 128
Walsh, Ellen, Irish immigrant, 128
Walsh, Michael and Stalker Wallis, 33
Walsh, Patrick, Irish day laborer, 128
Wapping, CT, see, South Windsor, CT 126
Wardwell, William, Connecticut River Company contractor, 162
Warehouse Point Bridge, 24
Warehouse Point, CT, East Windsor, 17, 36, 65, 74
Washington, D.C., 70
Water Quechy Falls, VT, 36
Watkinson, David, 61
Way, Peter, on job hazards to canal workers, 123
Welles, Martin, 98
West River, CT, 51
Wethersfield, CT, 16
White, Canvass, 48, 49, 64; and Edwin A. Douglass, 9, 35, 48, 49, 50, 64, 66, 70, 72, 93, 94, 96, 98 -109,112, 116, 118, 122, 143, 145, 146, 153, 154, 156, 162, 164; biography, 71 - 72; chief engineer of Connecticut River Company, 93; Enfield Falls engineer, 35; fee dispute with Connecticut River Company, 112 - 113
Whitney, El, 44, 45
Willey, Asa, 98
William Hall (steamer), 166
Norman Williams, secretary, Vermont Convention, 69
Williams, Samuel, paper mill owner, 174-175
William Weston, canal engineer, 38
Windsor Locks Canal; drawing of, 95
Windsor, CT, 16
  first bridge built, 23
Windsor Locks 65; bridges, 24
Windsor Locks Bridge, 23; canal specifications, 98
Windsor Locks Canal, 11,13
  industries along, 176 - 183
  purpose, 13; success a result of combined efforts, 188
Windsor Locks Canal Trail, 106
Windsor Locks, CT, 11; incorporated, 184; Irish labor camp in,121
Pine Meadow, 13
Winslow, Mary, wife of Elizur Lamberton, 160
Woburn, MA, 38
Wolcott, Erastus, petitions for river improvements, 25 - 26

Wolcott, Jonathan, 23
Wolcott's Ferry, 22
Woodbridge, Ward, 66
Woodley, Rev. R. D., visits Irish laborers, 12, 126 - 128

Woods, Asa Barr, lock tender, 158 - 159
World Trade Center, New York, NY, 126
Wright, Benjamin, 48, 55; hires Canvass While as assistant engineer, 72; hires Holmes Hutchinson as engineer, 71

ABOUT THE AUTHOR

J. Christopher Kervick is an instructor of local history at Mitchell College in New London, Connecticut. He recently completed three terms as First Selectman of his hometown of Windsor Locks, Connecticut, and, prior to that, served as a Connecticut Judge of Probate. He has been practicing law for over 36 years. A 1984 graduate of Fordham University, he is a 1987 graduate of Catholic University, Columbus School of Law.

His boyhood home was just three hundred yards west of the Windsor Locks Canal (originally known as the Enfield Falls Canal). In 2001, the State of Connecticut hired him to conduct a title search of the entire canal. The title search sparked his fascination with the story behind the canal, especially the plight of the over four hundred unidentified Irish immigrant laborers who built the canal, some of whom lost their lives in the effort.

Chris is married to Michele McElwaine Kervick, and they have three adult children, Katie, Mollie, and Dan. His poem, "A Canal Secret," has been published in Limerick, Ireland and has received widespread local appreciation.

Chris Kervick poses at St. Munchin's Church, Limerick, Ireland, where many of the Irish canal workers originated.

The Windsor Locks Canal was designed by The Connecticut Press of Madison, CT, and is composed in Adobe Garamond Pro 12/14 point typeface using Adobe InDesign, 20.0.1. Garamond is based on Roman types that were originally inspired by Aldus Manutius in 1495.

www.ingramcontent.com/pod-product-compliance
Lightning Source LLC
Chambersburg PA
CBHW070552010526
44118CB00012B/1297